CHARIOT
of FATE

Sigrun Norton

Copyright © 2023 Sigrun Norton
All rights reserved
First Edition

PAGE PUBLISHING
Conneaut Lake, PA

First originally published by Page Publishing 2023

ISBN 979-8-88654-141-0 (pbk)
ISBN 979-8-88654-147-2 (hc)
ISBN 979-8-88654-142-7 (digital)

Printed in the United States of America

I dedicate these Memoires to my two fabulous children, Roxanne and William who were such wonderful and welcome surprises in our lives and to my beautiful, handsome and clever Grandchildren, Emily, Rachel, Madison, Matthew and Taylor and the 4th Generation: Evelyn, Abigail, Wells and Brodie.

May they prosper in my pride.

I sit in silence, my mind at ease.
Pondering long and fateful years.
A childhood spent in war and peace with dolls and death, rich life to please.
Rubble and rolling hills combine in golden days
With terror sublime.

A new day blows a foreign horn, shrill and calming in one tone.
Move on it sings, succumb, and conquer
Move on and fashion a special tune
One just for you; so with luck's aim
You grow anew.

Life is vanishing in my rear-view mirror! Better write it down. This is a memoir of my life. It is not an objective, historical account with precise and accurate dates but rather a subjective story told from my perspective as best I remember.

I suppose all people consider themselves and their experiences unique. I feel that way only to a point. I also know I am one of many who were touched by fate or a higher force. We are called to follow a path not planned or expected.

I am asked frequently how I met my husband. My answer is my version of our unplanned meeting with unforeseen consequences for two people from different backgrounds and continents.

Chapter 1

DICHOTOMY OF FATE

From the age of twelve, I spent five years at an all-girls Ursuline Academy and Boarding School in Offenbach near Frankfurt on the River Main, Germany. In 1953, I was called north to my parents' home in Lingen/Ems, Germany, to finish high school there. I could have refused to leave the Ursuline's; however, it would have been a financial hardship for my parents. There was a medical emergency involved. When I arrived in the town where they now resided, I was only told that my dad had suffered a "breakdown." I never asked for details. I was young and egotistically immersed in my own little world.

I would not just be changing locations but also school systems from the State of Hesse to Lower Saxony; these states had totally different scholastic requirements. German high school education was demanding, but most difficult for me was changing to an extern student living at home rather than a boarder as I had been at the Ursuline Academy. Now here in the north, I lived a twenty-minute walk from the bus that transported students to a coveted high school, and the forty-minute bus ride to and from the school gave me terrible motion sickness. In Hessian high school system, we started learning the Latin language in third grade and French in the fifth grade. In Lower Saxony, it was the opposite. There were other differences that made a smooth adjustment almost impossible.

How I regretted my decision to leave the boarding school; it wore on my health. I had never before felt as forlorn as I did then, having to adopt an entirely new lifestyle. I spent one year with the Franciscan sisters traveling daily to their school. Then for the last year, I changed to a city-run public high school in Lingen that finished with an "Abitur" in economics. I never felt really comfortable living in my parents' downsized home they built in West Germany after escaping Polish-annexed Eastern Germany and being forced out of our temporary home in Czechoslovakia.

Upon graduation, I moved back to Frankfurt to take a position offered me by a friend of the family. I just did not feel like going to university at that time. I wanted to be independent. After the nearly six years spent in the Ursuline Convent and School in Offenbach, getting a job with a salary and moving into a rented room in Bockenheim, a mere forty minutes from my old boarding school, made Frankfurt feel almost like home.

I started training with a tourist company but found the money and the job lacking compared to my higher expectations. My enthusiasm waned quickly and decidedly. My only friend was the landlady who tried hard to entertain me. She had lost her husband and did not seem to have a close family. I craved more from life after a lengthy higher education. Every Sunday, I would flee to the Ursuline

Convent where Mater Benedikta, my aunt who was also the Mother Superior, welcomed me with open arms and good meals. This satisfied my need for companionship and food. I really was back at home.

One fine Sunday, about a month into this new life, I got ready for my weekly ritual and climbed into my chariot, a streetcar. I was bored and tired, and after paying the conductor, I fell asleep. My stop was supposed to be the last station, Offenbach, but I woke to find that I had been taken in completely the wrong direction. I was at Frankfurt's main railroad station, an unfamiliar place. Rather than taking another long trip back, I opted to listen to my growling stomach and headed for a nearby restaurant. There was only one table available in the establishment, and it situated me staring directly at both toilets. My thoroughly prudish education under the nuns had me looking for anything else to rest my eyes. I found another pair of eyes observing me from across the room. Looking back, I am amazed at all my unusual behavior.

At the time, I was just glad to look at something more pleasant than the toilets. My gaze was met by eyes that engaged mine with magic power. I felt unusually comforted and intrigued. I took a deep breath and relaxed. Actually, the word "relaxation" was not found in my vocabulary up to this moment. Upon seeing me, James allegedly told his friend, Tom, that if we could meet, he would marry me! He wasted no time, came to my table, and struck up a conversation in Swedish; he later told me because I was fair-haired. In his imagination, all Swedes were blond.

It did not take long for us to move into various other languages, threaded seamlessly by human chemistry. My

Swedish consisted of one pathetic sentence anyhow. James and another member of the air force hockey team were there looking for an absent colleague. My upbringing would have sounded an alarm had either one of the two men worn a uniform. We were not to speak to soldiers according to the nuns: "they would consider a young girl easy prey." But these two strangers were clad in interesting and colorful hockey casuals; surely, that was very different, said that little devilish voice in my brain.

My food arrived. Though I had been famished, I merely picked on it. Something strange had gotten a hold on me. Fate started to pull the strings. When James and Tom offered to escort me back to my rented room, I dropped my usual reserve and accepted a ride from not just one but two strangers. It was the start of a very long and interesting journey.

Chapter 2

OUT OF THE STARTING GATES

Ilsemarie, nee Kauffman and Wilhelm
Ferdinand Groeger at 1924 wedding.

I actually had four mothers: Ilsemarie who gave me life; Martha, my "nanny," a servant who filled it with love; Mater Benedikta, the Mother Superior of an Ursuline Convent in Offenbach/ Hesse, who was also my aunt, who programmed it; and Louise, in the United States, my mother-in-law's lifelong friend who was my kind and able guide into the new world.

Ilsemarie, my birth mother, was born in 1906 as the only daughter of a German textile merchant and his wife, Selma, who was of Hungarian ancestry. Her father, my grandfather, was a leader of his Silesian community and an elder in the local Lutheran Church. He also attained temporary fame when his face was printed on the inflation money used in early 1900. My grandmother, Selma, was an accomplished pianist. She was born in Glienecke, near Berlin, to a Hungarian chef serving Kaiser Wilhelm, the last German kaiser and king of Prussia who abdicated in 1918. My Hungarian-born great-grandfather was part of a so-called "Kitchen Nobility."

My father, Wilhelm, who went by the nickname "Panie," hailed from a family of German academics and businessmen. Wilhelm, oldest of six siblings, attended a military academy and studied law. He had two younger brothers and four younger sisters. The oldest girl, then called Agnes, joined the Ursuline Convent where her name was changed to Benedikta. The youngest of his sisters, Hildegard, was one of the earlier female medical students in Germany. Sadly, she contracted tuberculosis and died before finishing her medical training.

My parents married in 1924 after a very steamy relationship. Mother was then seventeen and my father twelve years older. They met when my grandfather hired my dad as a ski instructor. He "came, saw, and conquered" mother. He also saw a way to leave the "Freikorps Oberland" to join my grandfather in his already established, successful textile business.

House I was born in that My Grandfather built lower Silezia.

Grandfather Reinhold Kauffmann's hunting cottage in the garden next to the main house.

My mother and father already had three other children when I arrived. We lived in my grandparents' spacious home on the second floor. There were maids and other servants who worked in house and kitchen supporting both families. Mom still managed to spend plenty of time with her musical buddies hiking while my father pursued business and pleasure stretching from Silesia to Berlin and into the Austrian Alps. He met with his friends made both during WWI and in the "Freikorps Oberland." "Free corps" were formed by former military men after WWI. They were not considered an official army of the German nation. The 1919 "Treaty of Versailles" forbade Germany to rearm. Most of the well-known National Socialists (Nazis) hailed from these organizations and times. One man in particular was by my father's side in difficult times: "Sepp," Joseph Dietrich, who became Hitler's top SS general. These two became friends while serving in the First World War, my dad as an officer and the other as an enlisted man. Both joined the "Freecorps Oberland" in Bavaria after WWI.

So there I was, fourth daughter of Wilhelm and Ilsemarie, born three years before WWII enveloped Germany. Martha, my second mother figure, had worked in our family for some time, attending to everything in the house. She was the oldest of thirteen children raised on a farm where her father worked as a laborer. Martha was married to Leo, a tall, kind man employed as a forester by the count who owned a lot of land in Silesia and periodically occupied his local castle in my hometown.

The couple, Martha and Leo, had an apartment in town, but Martha's days were mostly spent working for my family. When I was born, she had recently had a miscarriage followed by a hysterectomy.

Martha, nurse/nanny, my second mother

As was common in those days, my mother gave birth at home in our large three-story manor situated between two streets and the Neisse River. Martel, as she was called, stood at the foot of the birthing bed, hoping my mother would hand me over into her care. Apart from her nursing duties, my mother obliged.

So I started my pampered, eventful, and colorful existence. Martha, my nanny, called me "Puppele" or "little doll" and loved me unconditionally. When I arrived, my grandfather had been dead for a number of years.

In addition to our parents, my three siblings and my widowed grandmother on mother's side lived in the main house. My grandmother retained an apartment on the first floor.

My dad was the new head of the family; he had a right-hand man who oversaw house and grounds but also was in charge of the automobiles. Some servants lived in the apartments located in separate buildings across the backyard, others appeared in the morning and disappeared in the evening.

In the following years after the start of WWII, my father rekindled friendships with buddies from WWI and the "Freikorps." Some of them, among them Adolf Hitler, had become the new rulers in Berlin. So my dad wanted to spend time in the capital. My mother and we children resided mainly on my mother's estate in Silesia.

While contracting to build the house in Berlin, my father lived with a Jewish couple, a professor and wife, who were related to the owners of the largest department store in the city of Berlin. Their son served with my father and gave his life for the fatherland in WWI.

Before joining the military, my dad studied law. With his legal background, he became a "syndikus" (a legal adviser) for the German Medical Society, and we moved into the finished house in the Nikolassee, a suburb of Berlin. A housekeeper and other personnel made it possible for the rest of us to live there intermittently, and while there, my older sisters attended school. My mother, however, still spent most of her time in our home in Silesia. While I was in my hometown and in Martha's care, life was smooth and pleasant.

In 1939, I was a plump, precocious three-year-old. WWII started for Germany. I can recall standing by a military truck being loaded with oddly shaped square bread. Up to this time, I remember bread as being round. Now everything suddenly changed; people were agitated around me, men were in uniforms, women were crying. There was general chaos and movement everywhere. I was whisked away to safety by my Martel.

Chapter 3

KATSCHER AND BERLIN: 1938–1942

House my Father built in Berlin.

House in Czechoslovakia.

In the years 1938–1942, we made important expansions to our family business and acquired other homes. My father, now running the textile firm by himself, moved some of the business operations to the newly annexed Sudetenland, part of Czechoslovakia, where the German-speaking population welcomed the Hitler administration and the jobs it brought with it.

A number of Czech firms were integrated into German manufacturing. Dad had always been a just thinker; he treated the Czech business owners with dignity, and the workers integrated into his textile company were paid life-sustaining salaries. This treatment of the Czechs would serve our family well but would not be sufficient to earn forgiveness for other employees like Raschter, the caretaker at our house in Katscher/Sudetenland.

This home was erected far out in the country. It provided its own electricity: a type of pool was situated at an elevated location, and a turbine at the lower trout-filled brook produced the power for the cottage. Cottage in name only, it was actually a modern two-story home decorated with frescoes by Pausewang, the foremost Silesian painter.

Even though it was a hiding place for many of my parents' Jewish friends, the house developed an unholy aura. I had a bad accident, getting badly burnt in a bath of hot water. One of my younger cousins drowned in the holding pond while her family lived there after fleeing from their own home during the Russian invasion, and one of our Czech employees was killed by an angry crowd.

When the Czech government reoccupied the Sudetenland in 1945, our caretaker, named Raschter, was punished because he worked for the Germans. A vengeful mob forced the fifty-year-old man from our house, tied him behind a horse, and dragged him to his death. This home was later requisitioned and occupied by a Czech family.

In our Berlin residence during those same years of 1938–1942, we had added an extensive air-raid bunker under our house. It was furnished with beds, food, and everything needed to sustain the many people who sought shelter there. When the bombing alarm sounded, the family and a lot of friends took cover under the Berlin home. One exception was always our housekeeper, Klara. She would not leave her room to join us in the cellar. There were friends, neighbors, and also strangers. We, children, had never seen who hurried down the steep stairs. Some were my parents' Jewish friends who spent more than just alarms hidden in our bomb shelter.

We, three youngest children, used to sleep in extra beds in the housekeeper's room during our visits to Berlin, which could be for a matter of days, weeks, or sometimes months. Klara, the housekeeper, would react to the alarm with the words: "If I have to die, I die in my bed!" Martel, my nanny and protector in Silesia, refused to travel to Berlin, so at four to six years of age, when I would visit in Berlin, I was left with different older caretakers. Klara and a frequently changing brigade of maids did the required jobs but wasted no love on me; sadly, on the days when the garbage men came in their gray uniforms, I was told they were looking for me. I was scared to death of these strange men and would escape to the neighbors.

Nikolassee was not just a place with villas and families. Across the street from our house stood a huge building in a wooded, parklike setting; it was "the Waldhaus," a psychiatric hospital where, I was told, they treated wounded "Stuka" pilots. These pilots would risk lives and limbs diving toward their target, delivering their bombs with deadly accuracy. Many suffered brain damage, and if they were lucky enough to live, they might end up in a sanitarium.

An older couple ran that hospital. When I was frightened or had a bad start to a day, they always welcomed me in their childless home. They were my refuge, serving as a substitute for my Martel back in Silesia. When I returned home in the evening, I never had the feeling I was missed.

Sundays in Berlin were different. My older sisters and I would get nicely dressed, and we would travel to visit Professor and Mrs. Mueller who had become friends when my father lived with them while he was building the Berlin home. At their house, we would eat cookies, play, and get lovely gifts when we won in any of the games we played during the afternoon. Here in the capital, we were first touched by the evils of the Nazi regime when we heard that Mrs. Mueller, who was Jewish, had been arrested. Her husband suffered a heart attack in the aftermath. Both vanished out of our lives into an abyss.

In the following months in the capital, a new person entered our family. My sister Ursula, then eighteen, got engaged to a German military officer. There was an elaborate wedding in 1942 after my sister, Krista, and I had departed Berlin. The bomb attacks got more and more frequent, and the "Christmas trees," phosphorus projectiles, fell closer to the house.

In 1942, I turned six and was ready to enter the first grade. I was sent back to my mother's estate in Silesia for safety and joined the first-grade elementary school in my hometown. What a relief it was to be back with my Martel, my nanny who had missed me and would now again protect me. From then on, we no longer went on our wonderful trips to Berlin and especially the Island of Rügen on the Ostsee. My father used to load us three younger children into his BMW convertible. We took off from Silesia without mother or my Martel, making frequent stops on the way. One such stop was a place with a very interesting sandbox filled with silica sand, connected to a glass-manufacturing plant. They sold glass objects; we purchased three chamber pots or, as we call them in German, "nocturnal urns." During the remainder of that trip, traveling in a convertible for all to see, you can imagine what we as four-, seven-, and ten-year-old children did: we wore our acquisitions as hats to the dismay of Papa and delight of onlookers.

When we arrived in Baabe on the Isle of Rügen, the real fun mostly began with a pillow fight between my two older sisters, Gisela and Krista. It developed into another embarrassment for the aunts and maids who were supposed to supervise us. Feathers flew everywhere; we were usually banished to the beach. Since I am exposing my dad's humiliating experiences, there were many in Berlin. Trying to be a good father, he treated us to a sail in his boat on the Wannsee. We managed a short distance before the wind increased, then all three of us started clamoring so loudly to get back on land that we attracted other boats who believed we needed rescuing to our thoroughly embarrassed father's chagrin. This one sad sailing adventure was both our maiden and final voyage on the Wannsee.

I, the youngest daughter, distinguished myself in another case. My sisters and I accompanied our father to the local bank in Nikolassee. He had a well-educated eye for beauty and spent a good deal of time enjoying chats with good-looking female bank employees. We were not happy waiting. With my sisters' urging, I stepped behind Papa, and my head being level with his hips, I used my only weapon: my teeth. I bit him unexpectedly and thoroughly in the rear. This remained a highlight of the day for us children, perhaps even of the rest of our banking adventures.

After Berlin, our 1942 arrival in Silesia was a very sobering event as my first-grade experience was unpleasant with an unkind male teacher doling out frequent and severe punishment for small infractions. His reign of terror even extended into the street. When we met him, we had to raise our right arm and shout "Heil Hitler." Noncompliance meant punishment next day in class.

Chapter 4

WAR YEARS: 1940+

As Germany expanded its war activity eastward, the entire country, even the small towns, felt the effect. Men were conscripted, and only the elderly, the underaged, the afflicted and those with essential occupations remained in towns and villages.

My father had wounds from WWI and was needed to run his textile mill. He and our factotum, who kept the house and cars running in spite of his epilepsy, were both among the men who stayed behind. Women ran businesses, and farmers' wives got help from foreign agricultural workers, many of whom, such as a number of Russians of western ancestry, were transferred from their German-occupied countries by force. While the farmers had to join the military, almost every farm in our area of Silesia employed a Ukrainian farmhand.

Fate for many families was cruel. I remember friends visiting our house, bringing news of their only sons, as young as eighteen, missing or killed in action on the eastern front. Our new family member, Hartmut, my oldest sister's husband, took charge of an army tank battalion and became a statistic, missing in action. We never saw him again. Ursula and her very small son, Jens, had never even left my parents' home and care.

And right in this chaotic time, my mother gave birth to a fifth daughter, born just before the war ended. Gabriele arrived at a broken world as a broken child with a heart condition. She was called a "blue baby" and died shortly after birth because there were no medicines or capable surgeons available for the civilian population. Beside our personal suffering, we watched injustice and pain afflict many living in my hometown. The war was consuming and affecting all of us. My father did his best to console, help, and save.

A family friend of Jewish ancestry, Alex, still a student at university, was to be shipped out to a so-called "Punishment Battalion" where the recruits were unarmed, digging trenches, and working for the German war effort without being able to defend themselves. Our father requested him as an essential worker for the textile mill. Alex lived to see the end of WWII working for Dad and was later able to finish his study of medicine. He remained faithful after 1945, helping my family through difficulties once the war was over.

Needing a bookkeeper, my father also rescued an Englishman from a concentration camp. His mother had begged for help. My father managed to extract him and put him to work in his textile business. How impressed I was to read in one of Alex's letters that he considered my dad a "hero" for standing up against the evils of the "Nazi regime."

In the year 1968 when we were stationed with the US Air Force in London, James and I happened to meet an English solicitor. He had made contact with the English spy's family and helped them claim my father's textile business as their own against all arrangements my dad had made during the war. The English solicitor's father had been the spy's handler. I learned that my father had been aware and supportive of the mission against the Nazi Regime, and he and the supposed "accountant" had made a pact during the war which would reward my dad's cooperation with the Allies after their victory.

The sounds of war came closer and consequently touched us constantly when one of my oldest sister's lifelong friends was conscripted and sent to the Russian front. He disappeared for many years. My family finally reconnected with him in 1975 in Washington, DC, where he was chief of "German Radio."

By the year 1944, our lives were drastically changed. We no longer had regular school sessions; most living male teachers were now soldiers. The school building housed Eastern Germans and Baltic allies fleeing from the westward Russian conquests. Our family took in various people. Among them were relatives and friends. My father's younger brother joined the military, and his wife, their eight-year-old daughter, and his father-in-law moved into one of the buildings behind our big house. A distant relative of my mother's, a displaced and impoverished nobleman and his housekeeper, took up residence in one of the extra bedrooms on the first floor of the main house. My mother would laughingly refer to him as "his excellency." In reality, both he and his faithful servant lived like paupers. Other relatives moved on to our country house in the still German-occupied Katscher/Sudetenland.

Since there was no school or structure for the children, we had to keep busy during the day when the adults had little time to spend with us. Apart from the attic, my favorite place to play was next to my parents' bedroom. There in the "gold room," my mother deposited all the unwanted frills my father banished, like silver vessels and crystal chandeliers.

When my grandparents lived on the second floor before my parents moved in, they had furnished this room with exquisite items including gold-painted chairs. I would slip in unnoticed and pretend to be a princess. Some days, I even managed to smuggle in a "crown" I found in the attic; it actually was a perforated, reticulated silver bowl. It fit my head perfectly. And princesses wore veils, didn't they? So an old lace tablecloth completed my regal garb.

Long after I was married in the US, my father brought this bowl along on one of his visits. It had been moved to Berlin during the war and was found there among the belongings that stayed in the house during the Allied invasion and occupation after the 1945 armistice. Our house remained in the American sector when Berlin was divided among the Allied victors. We were able to salvage some of the furniture and other content before selling the house in 1949.

The war progressed; we heard stories of atrocities committed by the Russians. We were told the Russian soldiers would rape and pillage wherever they advanced. Fear overwhelmed everyone, especially us children. Rumors and careless chatter by some older folks scared us to death. I conjured up images of severed body parts and heads.

My parents opposed Nazi philosophy, and my father suffered for his open resistance. He was incarcerated once during the war years but rescued then and many times after by his friend from WWI and Freikorps days who had served my father as an enlisted man in WWI and now in WWII became an SS general under the Hitler regime.

Fear is a powerful motivator, an enemy of all logic and reasoning; I prayed for German victory. Victory never came; God and the world had other plans. Hitler and his bride, Eva, committed suicide in their Berlin bunker. For a minute in time, we were told we had a new leader, Admiral Doenitz of the German Navy. Mockingly, we children raised our right arms and exclaimed "Heil Doenitz!"

Suddenly, the Second World War, with its horrible discoveries for the world and fateful consequences for my family, came to a crashing end. In February 1945, the Russian troops marched into my hometown. We heard the thumping of heavy boots while we huddled in the cellar of the main house next to the potatoes in winter storage and some of the coal we no longer used since my father had installed gas-fed heaters. This cellar had nothing in common with our well-stocked, specially built bunker under the house in Berlin. Here in this dank Silesian underground refuge were no beds, no food, no heat in winter; it was eerie, cold, and scary.

My mother had the bright idea that we might be safer outside. She, I, and my three older sisters, one with her infant in arms, dashed out into the garden toward the glasshouse used for early plantings.

No sooner had we reached that presumed hiding place than the first Russian soldier showed up, followed by two more. They chased us around the glass building; strangely enough, we all ran in one direction. It occurred to me that they were not really trying to catch us; it was a game for them, a tragicomedy.

We stopped running as a number of Russian officers emerged from the house and ran toward the men pursuing us. They were shouting in a strange language. I had a chance to look at the faces I feared so much. They appeared Asian, dark-skinned, and tired. There were some loud orders dispensed by two more well-attired Russian officers. Our adventure ended. We retreated back into our home. I heard that one of the maids the soldiers caught in the house was raped, but as a nine-year-old, I attached little significance to the word "rape."

Our home became headquarters for the Russian victors. To keep us safe and well-connected to doctors, my father arranged for our family to move into the local hospital. He stayed behind. It seemed the average Russian feared infectious diseases and would not invade a hospital. My mother and my oldest sister with her infant son went to maternity, and we, three younger sisters, were deposited in the children's ward. Martha and the other personnel stayed behind at our home.

What followed were boring days in medical care with forgettable food and endless strict discipline. I missed my Martel; my sisters missed their freedom.

About two weeks into our stay, my twelve-year-old sister, Krista, had enough of "safety" and decided to escape to the neighboring village where one of our aunts lived.

Of course, Krista, three years older than myself, needed reinforcements and chose me because I would follow directions, or in other words, I could be bullied into this six-mile walk.

We set out early one day after breakfast, packing some pieces of bread in a handkerchief. There was nothing to drink since small bottles did not exist or, at any rate, we didn't have any. No one would miss us before the noon meal. My other sister, Gisela, then fifteen, hoped we would get caught and punished, but aside from that, she was much too sophisticated to be bothered by her annoying siblings.

As Krista and I marched along the country road, a band of drunken Russian soldiers appeared in the distance, singing and shouting. We had heard the stories about their behavior, raping and ravaging any females nine to ninety. What it meant, I still did not know, but I wanted to run back to the hospital and safety. Krista, however, had her plans. She pushed me into the trench next to the road, face down in some comforting dirt. The men took a turn down another path, and I was pulled up by my sister to resume our escape.

The journey continued until we reached Lauterbach, the neighboring village, where our astounded aunt gave us food and shelter. Somehow, she got word to our parents. We spent the next few weeks with our aunt.

One of those days away from our family, we noticed a reddish glow in the evening sky. Much later, we were told that it had been the fires that consumed Dresden during the infamous Allied air raid that leveled the city and killed thousands of children waiting in railroad cars to be shipped out of the city to safety in the countryside.

Chapter 5

1945–1946

Months after the end of WWII, the Russians departed from Silesia and left Polish civilians in charge. Where there had been chaos, pure terror overwhelmed the German population. The Polish mayor took over our main house. We were literally kicked out, and all possessions were taken from us. One of the new occupants kicked my mother and pushed her hard enough to cause her to stagger through the upstairs' entrance foyer. She had always been a fair thinker who admired the Poles who immigrated to France and England. This ended her positive sentiment forever.

We, three younger children, lost most of our clothes; we especially missed the fur winter coats from the last lavish Christmas. We were too young to measure the value of life versus that of material possessions. In that moment in time, my parents and many others walked away from a world of comfortable existence. Life changed even more drastically. My family was ejected from the main house to a row of apartments across our backyard formerly occupied by servants or used as utility buildings. At one point during our stay there, we had thirty-six people in a two-bedroom space. Gallows humor carried the displaced grownups through the days and weeks that followed.

There were difficult and scary moments. We had a perfect view from our windows into a part of our old home, the big house where German soldiers, mostly young men, were interrogated. They were beaten and mistreated; we could hear their laments late into the night. One day, we saw a body being carried out and placed in a truck parked under our windows. Happenings like these heightened my anxiety, especially my concern about my nanny, Martha, who arrived in the morning and made her way home nightly. Germans had a curfew, and we all had to wear white identifying armbands. If Martha left a bit late, I would pray and stay awake late into the night; there were no telephones, therefore no way of knowing whether she was all right until she showed up the next day.

We had hardly any food. Waldemar, a Polish fellow who had been assigned by the mayor to manage the property for this family, took pity on us. He supplied us with the occasional just-butchered horse's head, with heart and lung still attached. My mother made stew out of the lung; this meal, "Luengli," is regarded as a delicacy in München, Germany; of course, that was made from beef. This horse meat became a boring but necessary staple for our family.

Sometimes, Waldemar brought other cuts of horse meat; apparently, there was a scarcity of cattle, so horses were slaughtered. My oldest sister, Ursula, an ardent equestrian, started to fret about our two horses still kept in their barn. Ulla was happy to see that Waldemar used them as transport animals. Ursula or "Ulla" had married a German tank commander named Hartmut while in Berlin. He was sent to the eastern front. My sister and her son never saw him again. Hartmut was MIA for more than ten years until he was officially declared deceased after she met her second husband-to-be.

We lived a miserable existence from day to day. We ate to survive. There were no more good days. Hope seemed to have died. We were happy when Waldemar showed up with bread and butter; he and his deliveries saved us from starvation.

It was winter, and all we had was a coal-burning stove and no wood or coal. My courageous twelve-year-old sister, Krista, and her crazy friends put their adventurous streak to use stealing coal at the railroad station to keep us from freezing. They would climb into the railcars at the station, exposing themselves to gunfire and risking the unknown and unimaginable consequences of capture.

Life progressed from bad to unbearable. My father who before had a reasonable understanding with the Russians was now in danger of imprisonment. I do not know why. The Polish government needed very little encouragement to take revenge on the occupied Germans. As a child, I was not aware how badly Polish people had been treated by their German captors. I might have realized; we were reaping what our government had sowed.

During that winter, Martha still lived in her apartment in town. For years, I would take my favorite dolls and toys along when she invited me to her home overnight. Her husband, Leo, who worked as a forester, would give up his marital privileges and move to the couch so I could enjoy life as an "only child." When Martha and Leo were ejected from their home, my dolls were left behind. We tried to retrieve them, but we were rejected; my tears and pleas were in vain. I lost twenty dolls, my children, in WWII. Leo was a tall very kind man in his forties who filled an essential forestry position. He was of Polish descent. That saved him from the military draft but sent him to his death in the Russian Gulags in Siberia following the armistice in 1945.

After a few weeks under our new foreign government, my father felt more than just threatened and decided to flee to his former business venue in Czechoslovakia. He pretended to go for a walk in the woods to collect mushrooms as he had done frequently. This time, he kept walking and crossed the border to a neighboring town, where he was taken in by one of the Czech friends, who now in turn helped him survive.

Dad left behind instructions for us children and Mother to follow as soon as we could. There were no official communications; we could only hope and ascertain later that he got across the Polish/ Czech boarder and did not get apprehended or killed by the Polish border patrol. A few days after Dad left, before his absence would be noticed, we, three younger siblings, started the same hike to the neighboring country, Czechoslovakia. We were accompanied by our former male house manager who suffered from epilepsy. Because of this condition, he was exempt from the draft and employed by our family.

The walk through dense woods were tricky and dangerous. If we were caught, the least punishment would be evacuation to some holding camp or expulsion as displaced persons to Germany. We had to leave in secrecy and in stages, so we children started the two- or three-hour adventure without our mother. At age ten, I was the youngest. I collapsed halfway to our destination because I was overdressed in three or four layers of clothing. I had to be carried back to rejoin my mother. My two older sisters, thirteen and sixteen, managed to reach and cross the border to the former "Protektorat," which was now again in Czech hands.

Gustav and I were dispatched again on the next day. This time, we succeeded. Unfortunately, on his return trip, he was apprehended, packed into one of the departing refugee railcars, and shipped to Germany. My mother never spoke to him and was left wondering about our well-being.

She and my oldest sister, Ursula, with her son, Jens, in a stroller were the last to escape. They had sewn many of their jewels into clothing. They could be sold to carry us through future uncertainties. It proved to be a wise move since only my older sisters were able to work and support the family. My father and mother had neither marketable skills nor did they speak the Czech language.

In Rokitnitz, we occupied a now-deserted villa from which my father's business manager and his family had been forcibly removed. They were shipped to a camp, and from there, they were evacuated to West Germany.

Martha had to remain in Mittelwalde, our hometown. She took care of my aging grandmother until both of them fell prey to the ongoing ethnic cleansing of Germans from now Polish-occupied Silesia to West Germany.

In Czechoslovakia, I suddenly had a real mother who was thrown by fate and misfortune out of her comfortable existence into strange surroundings and duties. We all adjusted poorly; I became obstinate and defiant, my mother irritable and impatient. I remember once when asked to obey, I sat down while my mother was chastising me; I then peed on the dining-room chair. I wasn't going to cry; unfortunately, I found different means of expressing my feelings. Mother was not used to dealing with me, and I did not know how to follow her orders. Martha had always been patient and never spoke a harsh word. I guess both my mother and I simply panicked.

During the day, I was safe in school, which I loved. Evenings and weekends, I avoided close contact with my mother. I prayed fervently for my Martha's return, but that never happened; I became very independent and self-sufficient. At age ten, I had graduated from being mothered.

There was a farm adjacent to our villa, now occupied by a German widow who had taken in a Czech laborer during the occupation. He became her partner and manager. My older sister, then sixteen and an animal lover, found employment on this farm. There were days when she suddenly appeared at home, unwilling to return to her labor, at times hysterically crying. My mother and father were helpless and clueless, never having been faced with such situations in their comfortable lives. There were shouting scenes, tears, and drama which I didn't understand until many years later when my sister confided that she had barely escaped being raped. Such incidents were common occurrences in her work environment. After this incident, Gisela, then sixteen, transferred to new employment at a local hotel. My sister evolved into a Cinderella without the happy ending. She never caught up on her education, never married. She was a pure victim of war.

My father, whose former business had been in this area, was rescued from boredom and evacuation by old Czech friends and employees. I remember biking to a family about six miles away; they stuffed our pockets with money to bring back to our parents. The government of this country was made up of exiles, and my dad discovered many acquaintances from his former working days. They helped him obtain sham papers to keep us from persecution and expulsion.

Chapter 6

1946: Rokitnitz, Lesany, and Liberec

Sometime toward the end of 1946, we were no longer tolerated as Germans in the town of Rokitnitz, Czech Socialist Republic, where my father's business had been located during the war. The British spy my father had extracted from a concentration camp and engaged in his business fled to West Germany with whatever money he could find in my father's former textile accounts. He left behind a mother, wife, and infant child. The grand plan of help and redemption my dad and the Englishman had arranged during the war was in shambles.

There was constant harassment by the Czech gendarmerie. Our family had many friends among the locals, but apparently, they also had enemies. Neither my mother nor my father had employment, which those in power required of everyone living in town, so there were frequent verification checks. The only one who had a place of work was my sixteen-year-old sister Gisela, and she was under attack by other forces as previously described.

In my dad's case, a somewhat comical way around the employment requirement came about. A truck driver formerly employed by our company, actually a Czech Freedom Fighter or "Partisan," now operated his own garage and repaired cars. When one verification check took place, Steiner dressed my father in stained company overalls and presented him to the inspectors. Everyone knew Dad did not know anything about repairing cars. Rousing laughter bought us another month's asylum.

One day, our luck and the benevolence of the new regime ran out. The country had been taken over by Communists. We had to report to the police department carrying only necessities; we were to be taken to a relocation camp. Men were separated from women and children. All were loaded into trains destined for a holding camp called Lesany.

Upon arrival, we were housed in separate barracks. We had to stand in line for all necessities, including food. The toilets were in outhouses; that was where the adults chatted daily, keeping up with news and rumors. A strange place for a moral boost but better than giving in to desperation. My mother recounted sitting "hind to hind" with a German professor from Prague University, separated only by a wooden partition. My parents and all other Germans in the camp were reduced to exchanging pleasantries in latrines by the new regime taking over the country.

A few weeks passed, then we once again climbed into a train destined for the new Germany. We made a stop in Liberec, a border town between the Czech Republic and Germany, and my family was suddenly taken off the train, loaded on a truck, and transported into the city by a young man named Karl. He spoke perfect German and Czech and was extremely persuasive and kind; he turned out to be a very close friend of my oldest sister, Ursula.

We had arrived in Liberec, a place where we would spend the next two and a half years. We found a beautiful house in Garden Street, only to be forced out by a bug infestation. My mother and Gisela were both awaking in agony from insect bites. For some reason, I wasn't on the insects' menu. I would have gladly remained there; it was an easy walk to my new elementary school and close to stores and life. The former occupants must have been removed without notice. It was fully equipped with everything we needed. But the bug infestation made it impossible for us to stay. We moved on to another abandoned house further out of town.

The only thing I happily gave up was the goat's milk my mother found at one of the farms close by the first house assigned to us. I was somewhat undernourished, and goat's milk has a high fat content and nutritional value, so I had to drink at least a half-liter of milk a day. I hated every drop. In addition to this horrible, strange-tasting milk, my mother dispensed fish oil with profuse praise and aplomb. When we finally found that house on the other side of town, the goat's milk was no longer available to my mother's chagrin and my delight.

The next home proved to be quite a find. It was a former bakery without bathrooms, so we had no bathtubs or showers. At least there were toilets on the main and upper floors. We decided to use the huge, wheeled dough-mixing vats as bathtubs. I cannot remember how we obtained warm water, but at least once a week, we had a veritable water park in the basement. Since there were no servants and my parents and older sisters had no real skills to live in these conditions, we led rather unconventional and disorderly lives; however, it was better than living in the former orderly home with bathrooms and bedbugs.

In the new location, my walk to school changed and grew extensively, but, of course, it was not just the walk to school that made life difficult. My mother and I did the grocery shopping. On most days, I picked up some things after school and carried a shopping bag with heavy bread plus a milk container in addition to my school satchel. The considerable load made me more than miserable, but I had no choice. The Czech shopkeepers would not sell my mother anything since it was obvious that she was German and spoke no Czech.

Money became a problem. The brooches, rings, and necklaces my mother and oldest sister had smuggled across the border in the lining of their clothes when we left our hometown in Silesia now really came in handy. The jewelry was sold one by one to provide food and clothing. I knew where that money was kept. There I was, eleven years old, with access to huge amounts of cash but with no sense of its value. I remember needing a coat and going out on my own to buy a sage green cashmere cloak for an enormous price. I had a twelve-year-old German friend who was in the same position; he would take money from the office safe at his family's former business now under Czech management. We, two kids, spent huge sums on entertainment for ourselves. It was a bizarre way of life, born out of the chaos and consequences of war.

On my walk to and from school, I would pass a field of poppies. When ripe, they turned into large pods filled with poppy seeds and made a welcome snack.

Chapter 7

THE SEEDY SIDE

School again became my refuge and delight, but the walk there and back was always a challenge. One day, an older boy who had been trailing me from a distance confronted me on my way home. He wanted me to come with him. My instincts awakened with a jolt. He looked quite menacing, a stick in his right hand, his left trying to grab my scrawny arm. As usual, I was carrying a metal milk container; it had a rounded handle attached with rivets that made it possible to swing the jug. With all my might and adrenaline giving me strength beyond my size, I struck the fellow in the knee with the milk spilling all over his shoes then pulled away and ran, dropping bread and milk.

I ran toward a house where I could get help; there a friend from school, who had gotten home long before me because she had no chores on her way home, called her parents. They calmed me with kind words. I went home and told no one about my adventure knowing my parents or sisters couldn't help or change anything. I was on my own and had better use my own intuition to stay safe. The boy who accosted me never showed up again, but another incident again showed me the seedy side of life.

One of my friends, Zdenka, used to take clothing and food to an old man, a friend of her family. One day, while I was at her house, she was asked to visit this family friend. I tagged along since I had nothing better to do. The man lived alone on the second floor of a dark apartment building. I walked up to the door with Zdenka, thinking I would be asked in. There was no rousing welcome for me; instead, the door opened just enough to admit my friend then shut with a resolute bang. My curiosity rose right along with the hairs on my neck. I waited a few minutes then looked through the keyhole. One could see a table and chairs near the door and part of a bed toward the left of the room.

I was as quiet as a mouse and could hear the man talking softly, pulling Zdenka in the left direction. I was naive to a fault, but there was this uncomfortable feeling that whatever was going to happen was not good. Zdenka kept objecting and gesturing toward the door. When the man pulled her onto the bed, an alarm went off in my head: *do something!* I started banging on the door with all my might, shouting loud enough that other tenants stuck their heads out of their apartments.

Then the door flew open, and Zdenka ran out, totally disheveled, with the old guy in pursuit. He was swinging a belt, trying to hit me. I was about half the size of my friend, who was two years older. I was also a fast runner and managed to skip down the stairs, leaving my friend and the man pursuing me far behind. I reached the front door, grabbed the handle, and swung it wide open, jumping onto the sidewalk nearly knocking over a young boy.

This was another episode never passed along to my parents. Zdenka apparently told her mother and no longer had to go to the tenement. My parents had no relationship with my friends, who were mainly Czech nationals, so they remained unaware of my exploits. As sad as it may sound, telling this story at home would have restricted my movements and friendships. I preferred to be free.

Czechoslovakia was a temporary place for us and for most Germans. We were only tolerated until the Communist takeover of 1948. The government installed in Prague after WWII included many patriots, such as Foreign Secretary Jan Masaryk, the son of the first president, Tomas Garrigue Masaryk. His reputation was besmirched by the new Communist administration. He allegedly committed suicide by jumping from a third-story window. Clearer heads believed that he was murdered to make way for the new leftist philosophy.

It wasn't long before my family and many other ethnic Germans still living in Czechoslovakia were notified of an impending transport to the Bohemian salt mines or, if lucky, to East Germany. While I attended school in Liberec, I adored the teachers at my elementary school; they taught me so much. All my love of literature, as well as my ability to read and write in other languages, stem from my time at that particular school. When I announced in class that my family was being moved along with other Germans, the teacher was speechless; from my school attendance and academic efforts, he never suspected my background. I wasn't sure whether to take the sentiment as a compliment or an insult.

Nevertheless, we were packed into another railroad car and sent on our way. Fall had arrived; the air was getting cool as the leaves slowly turned golden. Fate had its own plans for that particular train. The autumn rains overflowed rivers and flooded a bridge along our route to East Germany, so we werererouted toward Bavaria, part of West Germany. The train ride was primitive in cattle cars with zero comfort but lasted only about a day. We stopped at stations periodically to clean out the buckets that served as toilets.

In the fall of 1948, we arrived in Hof, West Germany, where we were processed with disinfecting spray, and the adults filled out mountains of paperwork about where they would live and whether they had relatives who would house them in West Germany. Children up to sixteen were put into a temporary children's home. This was my first experience totally away from my family. I adjusted well, even enjoyed the experience. It turned out to be the first of many separations.

One lesson that I had learned fast and well when I lost my dolls in Martel's apartment and also my nanny, Martel, was that no crying or begging would bring lost things back. That must have been what turned me into a "benign fatalist," believing there would always be something to substitute for my losses.

We moved on from this first stop in our new world. My parents and some friends went to Northern Germany. My two sisters, fifteen and eighteen, and I, age twelve, were left with my father's sister, Mater Benedikta. She was the Mother Superior of an Ursuline Academy in Offenbach in Hessen, West Germany. I was placed in the sexta, the sixth but actually the first grade of high school or gymnasium. One of the nuns, Mater Aloisia, became my personal tutor. My education had been interrupted in 1944 after two years in German elementary school. Many refugees had stopped in our town on their way west and were housed in the local schools. Our parents engaged some teachers in our homes; however, the tumultuous happenings of the time foiled any serious learning. During the stay in Czechoslovakia, I managed to get to fourth grade. I was immersed in that country's language. My German writing, my mother tongue and its intricacies, had been sorely neglected.

My sisters, too old for a start in a high school in Offenbach, were placed in the Ursuline's home-training program, which prepared young ladies to run a household. My older sister, fifteen, absolved her curriculum and departed for the town of Lingen (Ems) where my parents had found refuge. Somewhere near there, she started her career and life as a nurse for a family with a baby. My eighteen-year-old sister was being actively wooed and recruited by the nuns to join the Ursuline Convent. She never gave in. Gisela left after a practical education and found work and a life in Northern Germany and neighboring Holland.

Chapter 8

FATE FASHIONS THE GROUNDWORK

My tutor in sexta was a fountain of grace, beauty, and knowledge. Mater Aloisia, later changed to Sister Aloisia, got me reacquainted with the German language, speaking, and writing. She also coached me in other subjects where I was deficient. I gladly spent the extra hours after school learning grammar and style; I had a girlish crush on my tutor. One of my subjects in the first grade of this high school was English, a new language for me. Before that, there were two years of German and three years of Czech elementary education. There in Czech elementary school, students were also introduced to the Russian Cyrillic spelling. The other pupils of my new high school class had started in the spring, the beginning of the new school year. I arrived in the fall of 1948.

My desire to please Mater Aloisia, plus her persistent efforts to increase my knowledge, led to the opportunity for me to play the role of a child in an English-language production called "The Traveling Man," which was to be performed by the primas, the last two grades before the Abitur exam. I studied and learned my lines without understanding much of the English language. The play turned out well; it gave me a chance to shine, setting me up to skip the Quinta, the next grade after sexta, and catapulted me into the quarta, ironically the third grade of high school where I belonged chronologically as a twelve-year-old going on thirteen. The name "quarta" in Latin translates as "the fourth," but the system counts down rather than up; next come a lower and upper Tertia, lower and upper secunda (second); it ends with the prima, "the first" and final grade before the Abitur or graduation from high school.

The Ursuline teaching nuns had been driven out of Ratibor, Upper Silesia, which became a part of Poland after WWII. My aunt, Mater Benedikta, was Mother Superior of a convent and school. The Ursuline order took over a school building in Offenbach, West Germany, called Mary Ward that had formerly been run by a Protestant teaching order, the English ladies; now it was called St. Marien (St. Mary's) and became a state-approved Catholic high school and gymnasium that also had a boarding school for girls.

Until the school's residential villa was built, the first boarding school pupils lived in leased rooms in the neighborhood. My sisters, my cousin, and I lived in two rented rooms with a family close by.

This coveted semiprivate school was also attended by many girls from surrounding cities and villages. They would arrive and depart daily by trains and buses. Altogether, I consider that period of postwar life a microcosm of uprooted German society; the chaos of war brought people from many different areas and backgrounds together, all thirsty for learning and normalcy. One such peculiar character was the school's maintenance man or custodian; a nobleman and count uprooted and separated from his family and former wealth. This reality taught me the general wisdom to never judge a book by its cover.

Boarding school had a profound effect on my entire being and served to provide me with tools for coping with many changes in life. I learned that no boundaries can be placed on the mind. I used the newly gained imaginative powers and mechanisms to enjoy enormous pleasures and adventures without the potential physical dangers and disappointments that might have accompanied them in reality. My acquired knowledge helped me adjust to changing locations, occasions, and conditions.

To give an example, when hormones and feelings made me aware of boys, I chose as the object of my affections a young man who would peddle by the school gate daily. His name was Godfried, and he must have been about fifteen or sixteen years old. He was completely unaware of his important role in my world. I'm sure he never noticed the fourteen-year-old girl waiting for him at the gate of the Ursuline High School. I must also admit, had he stopped to speak to me, he most likely would have lost his assigned significance.

Though I wasn't yet aware of it, my West German school years established the basis for my future. The war had ravished the country; we were stripped of dignity and material goods. Now the American voluntary spirit touched our lives, laying the groundwork for a lifetime of admiration and gratitude to a special country and its exceptional people.

We were fed daily by the American religious group, the Quakers. A large truck would appear at the gate, and the lined-up throng of German high school girls would receive a wholesome meal with frills like chocolate, which was not to be found otherwise. "Care packages" of beautiful clothes arrived from the Ursuline Academies in the United States and made us the best-dressed schoolgirls around.

The Marshall Plan, named after George C. Marshall, lifted and revitalized the entire German economy. New life and substantial capital were pumped into Europe. All this undoubtedly played a role in my attraction to my future husband albeit subconsciously.

As I am writing, sitting in my place next to my son's house in Texas, watching the odd happenings of 2020 on television, I am forming my own opinion on the changes in this world. Starting with WWI, America was always the country that helped all over the globe, mainly through its faith-driven assistance programs and foreign aid and also through sacrificing its sons and daughters.

Over the years, the world has become reliant on this source of income and policing. The success of the European Union began with America's support. According to some economists, the EU, European Union, surpassed the gross national product and currency of the US in the year 2008.

(When a new spirit emerged in the US that wanted to invest in American infrastructure, stem domestic poverty, and reduce military intervention, many in the rest of the world apparently saw a great threat to the always-available source of aid.)

Chapter 9

The Ursuline Academy

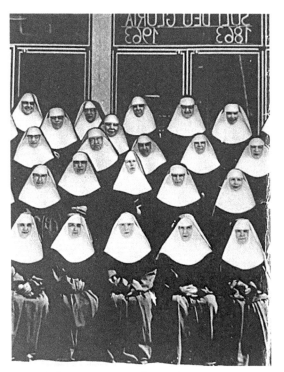

The Convent of Ursuline nuns.

The Mother Superior, Mater Benedikta. Third mother.

My life and academic knowledge expanded while I was in the Ursuline academy. I spent the weekdays in class with the girls, who, like me, lived in the villa across the street from the school building. I was the youngest of the five girls who shared one dormitory. We made our own beds, polished our shoes for Sunday Mass, and in general were expected to take care of our own space, belongings, and immediate surroundings. Once a week, though, one of the housekeeping sisters—who were not referred to as *Mater* or Mother as the teaching sisters were—cleaned the dorm thoroughly and changed our linens. Our clothes were kept clean by another group of work-assigned sisters. Besides the dormitory, this building also had a limited number of private and semiprivate rooms reserved for upper-class ladies.

One of those private rooms was occupied by a fascinating French girl named Therese. She socialized mostly with Mariele, her roommate, an Ursuline candidate, who was going to enter the convent after graduating with an "Abitur." Mariele wore a candidate's dress; it was black with an elbow-length cape and a rounded white collar. The small head cover allowed most of her beautiful brown hair to be visible. Therese, her roommate, however, wore the most fashionable dresses and suits, and her curly dark hair was adorned with colored bows and beautiful hairpins. She went out frequently, always carrying a gorgeous purse and a frilly French umbrella. Most memorable of all, she owned a number of huge perfume bottles.

We, younger girls, were thoroughly intrigued by Therese and somewhat envious of her. She would use splashes from her perfume bottles to reward small services, like fetching her tea or coffee. We were all eager to get close to her and admired her fascinating French accent which distinguished her grammatically correct German. Therese was the daughter of a French industrialist who lived in North Africa; he visited our school occasionally to see his daughter and to hone his grasp on the German language.

Mater Margaretha, the boarding school director, lived on the top floor of our villa; she was the head of the *Pensionat*. Mater Margaretha was our mother, judge, and supreme ruler. We all loved her obediently with a tinge of terror. She was the one, when you experienced your first menstrual period, would explain "the birds and the bees"; and she did it literally in those terms. From her stories, I deduced that getting too close to any male friend could put me in jeopardy. After my first intense kiss at a dance while visiting my parents up north, I worried for two weeks that I might be pregnant. I welcomed my menstrual cycle when it came to rescue me.

The weekends were wonderful. I spent them with my best friend, Luise, at her family's home about an hour from Offenbach. Her parents lived on a small farm run by her father and older brother. Her mother operated a mom-and-pop grocery store out of the family home. Luise was the youngest child with two siblings; her sister was ten years her senior and her brother was even older, so Luise was adored and spoiled by everyone in the house.

The little princess was sent to the Ursuline Academy for the best education in the area. I felt very lucky to be going home with her on weekends. I basked in her splendor. Those homey visits were a welcome substitute for life with my family hundreds of miles away. We built a friendship that is still thriving today. Other things in life were not totally uncomplicated. Sometimes, there were unexpected relationships and visitors, like one of my parents' friends, Hans (I called him Uncle), who must have been in his forties. He asked me to dinner, and I would take a train or streetcar to meet him in Frankfurt. Our outings were glamorous; I ate in the best restaurants and explored museums and parks. By lights out in the boarding school, I had to be back in my dorm room.

Hans enjoyed having me at his side, like a lovely young adornment that drew envious glances from strangers who, I suspect, assumed the worst of our relationship. I enjoyed the special attention, his interesting stories, and his generous gifts. He remained a complete, exceptionally good-looking benefactor and gentleman. Only once did he attempt to kiss me, and when I would not have it, he gave up like a chastised little boy. I looked forward to his visits as it gave a glimpse of outside and possible future life.

I was still in Offenbach in the year 1953, attending the *secunda* (the second, two years from finishing the *prima* with a diploma or *Abitur*), when I received an unexpected call from my mother. My father had suffered a nervous breakdown following some hardships in their new existence that were never explained to me. I was being called away from the place that had been my home since arriving in Germany. After we were driven from Czechoslovakia during the Communist takeover in the fall of 1948, my parents had found refuge in the town Lingen, Lower Saxony, located on the river Ems. What I did not realize until it was too late was how difficult the transfer would be. The school systems were very different from state to state in Germany. I was moving from Hesse to Lower Saxony in the North.

There was little I could do after getting the life-changing phone call from my mother about my father's medical condition. I informed my aunt and the school, received my transfer report card, and departed for Lingen/Ems right after the 1953 school year finished in the Ursuline High School in Offenbach/Hesse.

My arrival at home was less than pleasant since it all happened under the cloud of my father's illness and my mother's preoccupation with the problems that came with it. I was taken to another Maedchengymnasium in Thuine, about fifteen miles from Lingen, and introduced to the school director, a Franciscan nun, who welcomed me very warmly. She took note of my grades, was very complimentary, and wished me luck negotiating the difficult transfer. I returned to my family with an acceptance letter into the upper *secunda*, two grades before finishing with an Abitur in the prima.

Nothing went as expected. As described before, the difference between the school system I left in Hesse and the one I entered in Lower Saxony was crass and best described as unconquerable. I suffered through the 1953 school year earning passing grades to progress to the Prima. But I was not in my usual higher rank or spirits. Going on to study medicine, as my mother had hoped for me, was like a noose round my pride-injured neck. I wanted to pursue a career in art; painting was a passion.

My father, who was at home recuperating by now, was not too unhappy with this development and my indecisiveness. He told me that he had thought a path into business was always his hope for me. After all, he was a businessman. Art, literature, and music, my strong points were a "breadless cul-de-sac" to him. He actually called it "breadless art." I left the parochial high school system and joined a city- or government-run high school for business and economics.

Life became manageable, and I was again returned to my former elevated status. The curriculum taught was easy; all subjects taught were either new or my knowledge sufficient to fit in, and progress was smooth. I became class president, made many friends, and graduated a year later with a business degree in economics. I still did not wish to go on being a student. My life took a very unexpected turn when fate suddenly took over and sent me on a path no one in my family expected or any other close family members would take. I suddenly decided to go back to the only place that had been like a home to me since leaving Czechoslovakia in 1948. I returned to Frankfurt, close to my former boarding school. There, "fate" transported me on a fast-moving and kind cloud to another beautiful existence.

Chapter 10

A Leap of Fate

I met my husband three days before this image was taken. James and Sigrun. Park next to a tributary of the Rhine River.

Chapter 1 explained how I wound up in the restaurant where James, my future husband, sat with a friend, another air force hockey player. I had fallen asleep in the streetcar actually going in the wrong direction. I awakened at the Frankfurt Rail Station. My empty stomach did not allow another hour-long trip. I walked into the nearest food establishment. This is where James came into my life; fate intervened, took me by the hand, and I slipped out of my cocoon becoming a butterfly.

I found those mesmerizing eyes observing me from across the room. As I was still trying to gather my composure, these eyes grew into a young man moving toward me. He was a slim, medium, tall guy with a handsome face, dark curly hair, and he was dressed in a colorful silk jacket. A moment later, he sat down across from me, his eyes still fixed on mine. We started a somewhat awkward conversation. His first attempt was in an unfamiliar language, Swedish. Since I spoke hardly any Swedish, he did not get a response, just a smile from me. Still, I felt strangely engaged. Was it curiosity or interest that diffused my usual caution? A sense of comfort overcame me. To my surprise, there finally were words that I understood; the stranger spoke softly in an unfamiliar accented English. The young man introduced himself as "James" and stated he was Irish. Odd thoughts flashed through my hypnotized mind. I saw myself sitting by the sea mending nets. What a weird sensation while in a restaurant conversing with a stranger!
Karma had totally taken over; I was no longer the girl who walked in to order food. It appeared as if years passed in minutes. Food was delivered but got cold, and time passed swiftly as the stranger's friend came over to remind us that

it was time to leave. The two young men accompanied me back to my rented room in Frankfurt/Bockenheim. I agreed to see James again.

We did meet a few days later, and it felt like I had known him all my life. My English was just efficient enough to keep a conversation alive. Slowly, I perceived that his being from Ireland was somewhat far-fetched, that he actually was an American from Boston Massachusetts. His ancestors had immigrated from Ireland to the US. After only two more "dates," he actually asked whether I would consider venturing across the ocean to marry him. I accepted the proposal without hesitation.

But now I wanted to involve my aunt, Mother Superior of the Ursuline Convent. My new friend had no objections. The arrangements were made, and I took James there one Sunday afternoon. After pleasantries were exchanged, Mater Benedikta quizzed him about his family. The most important question for her concerned his religious affiliation. Luck was on my side. James was Catholic. He graduated from St. Charles, a Catholic high school near Boston which suited Mater Benedikta perfectly and sealed the "deal." My aunt's approval also gave me the feeling that it would simplify any further developments when we would involve my mother and father. Among the other guests at the convent that day was a friend of the Ursuline Convent, a cardinal of the Catholic Church. We received his blessing, and James was asked to drive him back to Frankfurt. This certainly was an unexpected honor. Acceptance of my future husband was assured.

A new chapter of my life had commenced. James totally dominated my thoughts and actions. We met after work almost daily. He got to know most of my friends, but I did not introduce him to my uncle who had provided my first job in Frankfurt. A few weeks into our friendship, we started to plan a trip to Lingen. I wanted to introduce James to my parents.

It was Memorial Day when we headed north for four hours by car to arrive in Lingen/Ems at my family's home at dinnertime. As I walked up to my father, my first words were "This is the man I am going to marry. Do not ask questions. He speaks very little German." My father's face seemed to turn pale, and Mother, standing at his side, had a forced smile. Both were polite and cordial as they were always with visitors. The conversation turned mainly to inquiries about Dad's sister, Mater Benedikta.

James spent the night at my aunt Erika's house just down the street, and I was back in my old room. We used that weekend to introduce my future husband to everyone, family and friends, including my former nanny, Martha, who was still around helping out everywhere. She told me that James looked like the gentleman I deserved. She liked him. He fit in the mix like he belonged. After her judgment, there were no more language difficulties. We communicated with smiles and gestures. James and I had acquired two gold rings in Frankfurt, anticipating approval by my parents. We placed the rings on our ring fingers on the left hand as customary in Germany and declared ourselves engaged. When we departed for Frankfurt, we felt our plans were cemented and our choices confirmed. We were an engaged couple.

My parents had announcements printed and sent them to family and friends. Back in Frankfurt at our jobs, it seemed like the days that separated us from each other were very tedious. There were no personal telephones, so we met every evening either in an Italian restaurant for an American-style dinner, or we visited the "King Bar" where many of James's American colleagues and friends spent their free time. Here I finally got to see men in air force uniforms. James always dressed in impeccably tailored suits, or he wore his hockey casuals.

One day, my uncle George invited me to his house, saying they planned a little party for me. My aunt took me aside into her bedroom so I could look at and borrow one of her gowns for that occasion. As we browsed through her well-stocked armoire, it occurred to me that I should mention James so he could be invited. I exclaimed, "Oh by way, I met this wonderful man, and we got engaged last weekend." In my naive mind, I expected applause and questions; however, there was a deafening silence. My aunt ran out of the bedroom looking for George then returned, confessing there was a friend, a Swiss business associate, they planned to introduce to me at the planned event. The plans for a party died rather quickly. I finally introduced James to Uncle George. Mid-August I decided to change jobs. I took a clerical position with an American company where I could practice and perfect my English.

We also spent some weekends with James's colleagues and their German girlfriends. I remember a pleasant girl named Frieda. She lived with her American boyfriend. One day, she confided that she was expecting a baby and hoped to be married soon. I was totally confused. Don't you first get married and have children later?

I waited in vain for the invitation to their wedding. That day for Frieda and John never came.

One day James informed me that Frieda had delivered a baby boy in one of the German hospitals. Of course, I bought some flowers and hurried to visit the new mom and child in hospital. My new friend seemed oddly sad. When I asked to see her baby, she made excuses. I finally left and resolved to visit when she and the child got home.

A couple of weeks later, I went to Frieda's apartment. It was customary to drop in on people in those days; announcing a visit was very rare and not expected. As I gained entrance, I found her in tears. Her American boyfriend was holding a bundle which he handed over to a strange young couple sitting in the living room who then rather hurriedly left the house and drove off in a waiting car. My presence was pretty much ignored and definitely not wanted. I departed feeling angry and very helpless. I never saw my friend again, but James provided an explanation that upset me even more. There were no real wedding plans; Frieda and John gave their child away.

This incomprehensible news caused a crisis in our own relationship. Suddenly, my heavenly new feelings about my fiancé became tinged with doubts and unwanted emotions. James tried in vain to tell me how he approached life, that he started with very different expectations of me. He took me in his arms while I cried unreasonably. All my anger about Frieda's situation exploded; fears and doubt that must have been present deep in my heart erupted like a geyser.

Again, as I had reacted with motion sickness when faced with my school change, my body reacted with an illness. I now developed shingles at an unusually young age of eighteen. It took some time for me to recover. James remained very kind, patient, and protective, even accompanying me to see my doctor who took a role as counselor, telling me I was fortunate to have such a dedicated fiancé. Medicine worked to heal my malady.

James was even more attentive than before; he brought unusual gifts. For instance, he presented me with a jar of Pond's Cold Creme instead of the usual flowers. He worked very hard to erase the horrible memory of Frieda and her child's fate. His obvious dedication and love for me restored my trust in our future.

Autumn descended. James's departure at the end of October was in sight. We still met frequently for dinners but avoided contact with other Americans since it seemed to upset me and bring back questions about Frieda and her fate. One day, James suggested that we take a road trip to Alsace, France. I packed a small travel bag for a long weekend, and when I saw James's car pulling up in front of my temporary home in Bockenheim, I ran down and jumped in as if I were fleeing from there forever.

We drove off Friday afternoon to arrive at our destination at dinnertime. It was easy to find a lovely restaurant directly in the City of Strasbourg. We picked one that had convenient street parking. The meal was as delicious and, as expected in this location, hideously expensive. It was dark outside when we checked into a hotel. Just then, I vividly perceived that the two of us were going to stay in one room overnight. James had visited my rented abode in Frankfurt only during the day. The landlady was usually at home. He was polite to her face but resented her presence. When we were alone, he called her by horrible names like "the old witch" or "old bitch." The latter was dropped fast because I asked for an explanation since the word was nowhere in my dictionary as a synonym for an old woman.

As James registered our names with the receptionist, I waved to the lady at the hotel desk, prominently showing my ring. I did feel somehow awkward. She just gave me her professional, superficial smile in return. We received the keys and walked down the long corridor to our room and entered. There we found a bottle of red wine on the table. We sat down in the two chairs provided, and I studied the rest of the furniture consisting of an armoire and two beds. A soothing atmosphere, and the wine carried us back to our first moments together until we found ourselves on one of the beds. Now my overactive fantasy conjured up pictures from literature imprinted in my mind. Frieda, the American sergeant's girlfriend, appeared to me in the frightening form of "Gretchen from Goethe's *Faust*." Magic and I fell apart. I suddenly panicked. James was totally startled. Though he was well-educated, he had not read Goethe, much less *Faust*. He could not imagine what was going on in my overactive mind. We talked for a while; I tried to explain my behavior, then we slipped into our separate beds.

Next day, we departed for Switzerland. In Zurich, we spent the day sightseeing, had a fabulous meal featuring the famous Swiss truffle mushrooms, and then finally stopped for the night at the "Hotel du Theatre." We were extremely tired when we entered our room. Again, there was that bottle of wine from heaven or, perhaps, from hell. The atmosphere was much more romantic than the night before. No ghosts invaded my mind. We were together and overwhelmed with emotions. My resistance waned steadily, and it was shed with our clothes; embers grew into passionate flames long overdue, and love prevailed. Frieda or "Gretchen" was forever banished from our lives, lives we vowed to spend together until death would us part.

We returned to Frankfurt determined to set in motion all the plans toward James's separation from the air force and for my voyage into a new married life and new world. We were not quite sure how my family would view my departure, so James bought an open ticket for me to travel to New York on the SS *America*. I quit my job in Frankfurt and applied for a visa to enter the United States.

In the meantime, until the visa was granted, I went to spend the months before my ocean voyage with my parents in Lingen/Ems. My mother and Martha, my former nanny, took this opportunity to teach me some basics in cooking and running a household since that had definitely not been in the curriculum of any of my schooling. Now I realized what the great planner, James, had in mind persuading me to return home.

The days crawled. Patience was never my virtue. We did not hear anything about my entry into the United States for weeks. My father drove me to the American consulate in Bremen. There I was told that they were processing stacks of visa applications. It might be many months before my entry visa would be issued to let me travel and be reunited with the man waiting for me. I felt very anonymous and frustrated.

My next letter to James, and there were many going to and coming from him, expressed my feelings of helplessness. James was living with his mother in Waltham, Massachusetts. He discovered a connection somewhere in the Norton family to a senator in Washington, DC, and so that family member was contacted. Someone promptly examined my application to enter the United States. Within a couple of weeks, the Nortons received a telegram from the senator that my visa had been processed and sent to me. To us in Germany, it seemed like a miracle. I felt special again, not like a piece of paper in a "stack." My family and I started preparing for my departure shortly before Christmas 1955. The holiday itself was to be spent on board the ship, SS *America*, which was to travel from Bremerhaven, Germany, to Southampton, England, and on to New York City in the United States, my future home.

Chapter 11

A Colorful Voyage

A colorful but lonely autumn turned into winter, signaling my final days with my German family. I was sad one minute, and the next, my heart seemed to skip a beat when I thought of James awaiting me in that wonderful new world. My old nanny, Martel, tugged on my sleeve asking why I was smiling. After all, I was traveling into the unknown, leaving everyone and everything familiar behind. I told her that I am putting my life and happiness into the kind and capable hands of the man she met when I brought him to meet her and the family. She smiled, stroked my hair as I bent toward her small shrunken body, and said, "You know best, Puppele!" I knew as long as she breathed, she would love me wherever I was. This strong bond was mutual!

Lots of friends and relatives arrived to wish me farewell on the day my father loaded my two suitcases in his car to take me to the ship by the same name as my destination, the (SS) "*America.*" The last day at home in Germany, only my body moved between the crowd gathered to say goodbye. I took all the good wishes with proper thanks and smiles. My spirit and heart were already on their way with wings that hardly needed a steamship. After all, I was going to start a new life with James.

My mother's sad face and Nanny Martel's tears as we departed did not touch me as they would have on other occasions. I was steeped in joy and anticipation to finally join the man who had changed my world.

I got in my dad's car and was waving happily when my mother said, "Travel safely, write, and remember, we'll keep the bed made for you, just in case." I grinned and thought, *Just in case of* what? I was totally convinced this was the right decision for me. My father started the car; we drove away. The people waving became smaller and smaller, and when we turned onto the main road, they were gone.

We hardly spoke during the two-hour drive. Both of us were trying not to become too emotional. The car radio played pleasant Christmas music and diverted our attention. My father was occupied finding his way to the place where the vessel that was to carry me to the new continent was docked. We finally arrived in Bremerhaven where my ship was moored. My dad boarded the ship with me. He was lugging my suitcases. I followed carrying my train case, a large handbag, and a French umbrella, a similar one Therese, the French girl in Offenbach, had and I dreamed of since my boarding school days. I was dressed in a gray fashionable suit and, of course, wore a hat, gloves, and high-heeled shoes like a proper lady. I finally looked very much like my idol, Therese. A ship's employee took my bags. My father hugged me and left while I followed a steward to the cabin shared with another German girl.

After a perfunctory introduction to the other passenger who was already in the cabin, I became so busy taking in my surroundings that I cannot even remember what my roommate looked like. I paid no attention to her name either. I did not even go up to wave farewell from the railing as the ship pulled away from the dock. My father had said he would not linger; it made parting less painful. I loved my cheery father very much and always felt pity for him in his diminished status after the war ended. Everything my grandfather had achieved in business and my father took over and expanded was taken from us when our homeland was divided after the WWII armistice.

My mother became negative and bitter. She would never except our fate as "exiled and displaced Germans" from Silesia, her home since birth, which was annexed to Poland. We were not considered refugees. Like other Germans from Silesia and adjacent areas as the former "Sudetenland" under the Hitler regime, we were called "displaced." Actually, we had been removed from our homes or expelled by "ethnic cleansing." My father got a paltry 4000 "Reichsmark," German currency as "restitution" for a fortune lost. My parents had to adjust and attempt to live on Dad's officer's pension from WWI when they finally settled in West Germany. We still owned a furnished house in the Western Zone of Berlin, and my father hoped to sell it if and when possible. Now, with my choice of husband, they were losing their youngest daughter to a faraway country. I resolved to stay closely connected with my family and friends in Europe.

Alone on the ship, I searched for the cabin-class dining room since there were three classes with their specific areas.

I found some friendly faces at a table and ordered food. All of us sitting there were lost in thoughts, so we conversed sparingly. The waiter reminded us to return to the same table for breakfast. I finished my meal and hurried to my cabin where I undressed, got in bed, and fell asleep immediately.

In the morning, there was a knock on our door. The other girl jumped up and grabbed a tray with two cups of tea brought in by the steward. We drank the tea, got dressed, after separately using the bathroom, and went upstairs to the same table we had found on the day of our arrival. Breakfast was good and almost too much for any moderate appetite. I left earlier than my tablemates to explore the ship. I was a planner even then and knew we would celebrate Christmas on board. How could I find a chapel to attend church services over the holy days? The steward proved to be a treasure chest of knowledge and help. Perhaps he shared the same belief and faith? "Come to the first-class area," I was told, and he would make sure I got into the chapel for the religious service. Actually, he found the schedule for me and picked me up at the cabin door.

The two days before Christmas though were quite awful. We sailed into bad weather; the sea looked like a boiling pot of water. The other girl in the cabin and I got terribly seasick. No pills or potions helped. I vomited an entire day and could not eat for two days. We were brought tea, and someone cleaned incessantly. The ship's doctor assured both of us we would live to see Christmas. He was right.

The third day of our voyage was a charm. All illness vanished miraculously. We felt well, and I set out on my own to find entertainment. Somewhere in one of the lounges, I discovered a table with people debating. Never having been shy, I took an available chair and joined in the discussion. One middle-aged man handed me a calling card with his name and occupation as president of a dairy farm situated near Lake Erie. Two other young army lieutenants from North Carolina were returning home to become civilians. One of them wrote down his address, and I stuck it in my purse. His name was John, and he was returning to a North Carolina Furniture Enterprise. From that day on, I would eat at my chosen dining table and then search out the people I had met in the friendly conversational group. Debating was my forte; I had always been part of a debating team in high school. Here I found a practical application of my theoretical learning.

Christmas brought beautiful decorations and fabulous meals with cookies and deserts. The sea looked like a mirror, and the grand ship sailed smoothly. I participated in the first-class chapel services and spent the rest of the day grazing the tables with sweet delicacies and talking to pleasant strangers in the first-class lounge. Someone mentioned that a famous actress named Ginger Rogers was sailing with us. Of course, the name meant nothing to me. Even if I saw her, I could not have reacted. We rarely watched foreign or any movies. I couldn't have been happier with all the ship's accommodations. We had baskets of fruit and goodies brought into our cabin, and all our needs were met. After a five-day journey over Christmas 1955, the SS *America* dropped anchor in New York City Harbor.

I was busy packing when the steward knocked and entered the cabin. "I hope you are pleased to hear this," he addressed me. "There will be interviews of some passengers from the different fare classes on this ship, and you were chosen to represent the cabin class." I was thrilled. New world, new life, and new excitement. I finished packing, grabbed my hat, purse, and the cherished French umbrella then followed the steward to the top deck.

There was a large couch in one of the ship's lounge areas. A middle-aged man with a strangely shaped mustache was already there answering questions posed by a male reporter. I sat down at the other end of the sofa listening attentively. This impressive-looking passenger was obviously representing the first class. The man being interviewed looked very imposing, had a strong foreign accent, and called himself Salvador Dali. I heard that he was an artist coming from Spain traveling on to California. He told a lengthy and highly interesting story but did not acknowledge me in any way. I thought he certainly was arrogant if not a bit rude. His interview ended, and the reporter's attention turned to me. I was expecting lots of questions, but the newspaper guy did not really want to hear much from me. He asked me to come along to the stairs in the interior of the ship where he took photographs of me with or without my hat, striking unusual poses holding my umbrella. What a disappointment for me; he was not interested in who I was, where I came from, and why I was here. He just snapped numerous pictures.

I had to drop my musing; the steward appeared carrying my baggage. He handed it to a porter, made an elegant bow, and urged me to join the disembarking throng.

I searched the crowd below the gangway for a known face. There he was, a smiling, waving James with a gray-haired lady at his side. I flew down the stairs into his arms for a long and loving hug as the lady tipped the porter. I believe I said "hello" to her, making a curtsy. She looked friendly but a little severe, more like one of my former teachers.

We moved away from the SS *America*. James grabbed my belongings to find his car in a garage. The gray-haired woman was Helen Norton, my future mother-in-law. James drove the car. "Ma Helen," as she asked me to call her, sat in the right passenger seat, and I crawled happily in the back. The same feeling of security experienced when first meeting James in Frankfurt took a hold of me. I sank, with an audible sigh of relief, into the comfortable back seat of a black Chevy Coupe.

We drove toward Waltham, near Boston, stopping only once to buy some fuel and to have a bathroom break. Those bathrooms or toilets were my first comparison to Germany's customs on the *autobahn*! What? No money needed? It seemed very civil not to charge for basic human needs!

When we arrived at the Norton home on High Street in Waltham about four hours later, we found a full house. James's two older brothers, Bob and Thomas, and his youngest sister, Mary Lou, came out to welcome us. I knew there was a fourth son, Billy, in the army, who could not be at home at that time. The youngest brother, Walter, was in the living room but did not even take his long scrawny legs off the low living room table as we walked in. He kept watching something on a television set and barely waved hello. Then I noticed another striking lady sizing me up.

She was very tall, conservatively dressed, and had a pleasant face smiling at me with large healthy teeth. This was Louise; I recognized her from stories James had told me about his mother's childhood best friend. They had known each other since "kindergarten."

My American Mother Louise Neilon

I could hardly believe my ears; from now on, James and I were to be separated, the family had decided. I was moving into Louise's home, he was staying at his family compound, consisting of three homes built where two roads, High Street and Cedar Street, met. At nightfall, Louise took me and my luggage to her large Victorian home. She was very fond of Europeans and over the years had taken in a number of Germans on their way to new lives in the US. I felt I had arrived in a prepared nest provided by a wonderful large and caring family.

James and I thought we would simply get married by a judge, but Ma Helen and Louise would not hear of it. They hatched big plans. There was to be a wedding at St. Charles Church in Waltham with all the usual pomp, relatives, and friends present. Unfortunately, my German family had not expected these elaborate plans. They were unable to attend. The weeks before our wedding were like a form of torture for James and me. We had weekends to ourselves, but there seemed always to be someone present to prevent any privacy between us.

During the week in the time leading up to our wedding, James drove to his future place of employment for training in Concord, New Hampshire. He had been hired by the US government as an air traffic controller.

When he visited Louise's house, she seemed like a Cerberus, the three-headed monster guarding Hades. She was to keep us from being alone to prevent any privacy between us. She insisted that we should not spend time in my bedroom before the marriage. Both of us accepted the house rules reluctantly. We managed to drive to one of the parks near Waltham, stealing a few hugs and kisses in the car. But it was winter, and Norumbega Park was cold and unwelcoming. James tried one of his old customs, taking me to a place called "Trapelo Rd." I recall he said that was where students used to go "necking," a new word for my vocabulary. There I would spoil his plans, not being used to and not approving of such activity. Reality was never far enough for us to relax.

The wedding day was set to 26 January 1956, a month after my arrival. It seemed like an eternity waiting one more month. Other preparations kept us busy. First order before a marriage in those days was a blood test. It screened for venereal and genetic diseases. *Why on earth?* I thought, but I went along with all requests to finally have James to myself again. We also had to see the cleric for "premarriage counseling." I hesitantly accompanied James to St. Charles Church Rectory. When the pastor breached the subject of birth control, I snickered. Actually, it was very embarrassing that a stranger would get so personal; however, I was prepared to endure everything to facilitate our marriage.

From these necessities, we went on to the most pleasant task—namely, finding a wedding gown. Ma Helen took me to the main department store, "Grover Cronin's," on Moody Street in Waltham. I spied a lovely wedding dress made of alençon lace which, without a doubt, was the most expensive one in the shop. Ma Helen bought it for me. Mary Lou, James's sister, was to be my maid of honor; for her, we ordered a special bridesmaid's dress. James was notified to report to work at the CAA, Civil Aeronautic Association (now FAA, Federal Aviation Association), in Concord, New Hampshire, earlier than expected. Our wedding date had to be changed to the twenty-second of January. Oh, one week sooner made us very happy. All plans progressed without problems.

Everything was beautiful on our big day. Since my family, especially my father, was not present, Bob, the oldest of James's brothers, lead me to the altar, or as the others called it, gave me away. Bob lifted my somber, dreamlike mood by whispering witty remarks as we proceeded down the aisle. My family let me know later that Bobby had written a most beautiful and respectful letter to my father, thanking him for the honor of standing in for him at the wedding. He remained my favorite brother-in-law to this day. Another Norton relative, Margaret, graced the ceremony with her lovely voice. The new families had thought of everything, and I was eager to join them.

The wedding reception was held in Louise's spacious home. The dining room table where we cut our wedding cake is still with me and reminds me daily of one the most important days in our lives. Louise told me many years after our wedding that she "adopted" me the minute we met. She turned out to be my fourth "mother" besides Ilsemarie, the mother who gave me life, Martha, my nanny who loved me like her own, and an aunt, Mater Benedikta of the Ursulines, who loved my soul.

Chapter 12

A WEDDING WITH CONSEQUENCE

Day James and Sigrun got married in St. Charles (January 22, 1956).

For James and me, Saturday, 22 January 1956, was the second most fateful day next to the unforeseen meeting in Frankfurt, Germany. This date was our wedding day in Waltham, Massachusetts. The month we had been separated seemed like a decade. During the church ceremony, I was in a dreamlike trance, fulfilling all tasks asked of me, uttering all words elicited by the pastor sanctioning our vows. Time stood still for the entire ceremony and during the reception at Townsend Street, Louise's home. There were pictures and films taken in the church. I only saw oblivious beings who moved like marionettes in a theater. The beautiful music and songs coming from the organ accompanied by the voice of a gifted relative lifted our hearts and everyone's spirits.

We walked on clouds after again officially exchanging our "engagement" rings bought in Germany. We had worn them since the engagement, 29 May 1955. They carried each other's initials. As we left the church with applause, we were aiming toward our well-attended wedding reception at Louise's house. There we bowed to traditions, receiving guests and cutting the cake, but my mind, and by the looks I caught from my new husband, his also were preoccupied with thoughts of our "honeymoon" getaway, an apartment James had rented while on his work-related training jaunts in Concord, New Hampshire.

The wedding celebration came to an end. We hastily changed into our travel clothes. The family members and friends present showered us with rice and good wishes, then the black Chevrolet Coupe loaded with most of our belongings, carried us to the home we would finally share in well-deserved privacy. We arrived in the City of Concord, New Hampshire, early enough on Saturday evening to see the stately house inviting us to spend the first night as a married couple. The furnished flat spanned half of the top floor in a proper mansion, the Peasley Funeral Home. James, the little devil, knew I would not be offended or frightened to live there since we had no so called "funeral homes" in my home country of Germany. For me, it was merely a beautiful big house. James had prepared a first home for us in a grand manor in Concord, New Hampshire's capital. I was impressed; this building seemed even larger than Louise's expansive Victorian house which had been my home for a month.

We climbed to the third floor where Mrs. Peasley, our new landlady, welcomed us warmly. The apartment was totally furnished, including dishes, towels, and linens. The double bed behind a glass door looked inviting, and there was a pleasant smell of coffee wafting from the kitchen. Mrs. Peasley showed us around and then excused herself, and we were alone with our pounding hearts. I felt the warmth already experienced at my first encounter with James but added was an overwhelming tenor of expectation tinged with a little bit of apprehension and confusion. How did we get this far; what would the future and fate bring?

It was my new husband who seized the moment with a mundane sentence: "Let's have a cup of this coffee." That remark soothed my trepidations and dimmed any tension. To me, it felt like a heavy weight was lifted from body and mind. The pressure of the past months turned into relief, hope, and euphoria. James's hazel-colored eyes were smiling again in such a seductive way. My composure melted, and we abandoned coffee and kitchen, danced through the glass door leading to the bedroom and into a blissful night. Clothes flew off, and the last thing I remember was a voice calling me "Mrs. Norton" and thinking, *I am at home. Life will be beautiful!*

Morning awakened us with blue skies and frigid January temperatures. What did it matter, we had each other to keep us cuddly warm. From now on, only we determined our wide-open future. But in spite of all the freedom, a disciplined religious upbringing, and, of course, the homey church bells gave us the upcoming Sunday compass. We dressed in the same formal clothes we had so hurriedly discarded the evening before and prepared for our first visit to St. John's Church just steps away from our new home on Main Street.

The service was celebrated like in any other church. However, I noticed a difference in the greeting by the officiating cleric as we were leaving. This young man held out his hand as if he actually was glad to see us. We introduced ourselves proudly as Mr. and Mrs. Norton then proceeded to the rectory to make that official. We were happy to be a part of our common faith and family. By now, it was lunchtime, but we rushed home to enjoy our new status instead of joining friends and James's colleagues at a local restaurant.

Lunch consisted of two homemade grilled cheese sandwiches and tea, and desert was priceless.

Sunday was ours; we spent it savoring every minute with each other without interruption or care in this world. The night was full of tenderness and promises. Monday morning called us back into reality; James had to go to his job at the Concord Airport. I would spend my first day as a housewife waiting for a husband to return from work. Making beds was familiar and easy; boarding school had taught me well. Cooking my first meal, however, was worse than passing a school exam. Even though my mother and my nanny had tried to domesticate me in the two months spent at home with my family before departing for the United States, I floundered just reheating the beef stew Louise had packed for our first meal in Concord. Handling a gas stove was not part of my training either, so I burnt the stew, undercooked the potatoes, and forgot to turn on the burner to warm the carrots. I was very frustrated, close to tears when James appeared shortly after four o'clock.

I found it hard to believe, but he walked in, looked around, and almost died laughing. *What a totally unexpected reaction*, I thought. He turned off the stove, gave me a huge hug, and said, "Did you really think I married you for your cooking or cleaning skills?" He didn't look disappointed, unhappy, or hungry; he seemed amused, almost happy to come to my rescue. I thought I can live with that attitude and relaxed. We ate some of the food, threw the rest in the trash can, and cuddled on the comfortable sofa in the living room until it was time to retire behind the inviting glass door.

All these "firsts" were welcome adventures. We thoroughly enjoyed each other's company, planning meals for a week; everything seemed better than we could have imagined. It was amazing how we agreed on so many tastes, likes, and levels we had never explored before. I compared our plans to those of some of the other Americans and their girlfriends we had met in Frankfurt, Germany. Many of these relationships did not survive.

It was mainly James who had the foresight to postpone our marriage, wanting me to travel to the US to get to know his family. He had asked his mother to be my sponsor; that almost guaranteed my acceptance. I felt very lucky; though my "knight in shining armor" was only four years older than myself, his good planning made him a wise man.

A few weeks after our move to Concord, we made a return visit to family and friends in Waltham, Massachusetts. Louise now accommodated both of us in "my" formerly assigned bedroom. We became her family. Ma Helen felt slighted because we spent most of our time in Louise's home that she stopped considering Louise her friend. Both women did not speak to each other for many years.

There were so many things to do and people to visit in Massachusetts that a weekend never seemed long enough. We sought out Peggy and Joe, cousins, who were closest in age. They had gone to school with James. Joe grew up in the same family compound in one of the three houses built for the extended family. We felt drawn to these cousins; we loved their company. Perhaps we were looking for a little guidance mixed in with the entertainment they provided. Peggy and Joe already had an infant son, Joey; their lives could be a mirror of our future. I adored them; they were beautiful people living desirable, well-managed lives.

Winter in New Hampshire will always have a special place in my heart. The temperatures and the snowfall were extreme. Some clown at the airport had impaled a tennis ball on our Chevy's extended antenna. I was annoyed but learned the significance of this presumed prank one workday morning after it snowed night and day. Our car parked in the street in front of the Peasley's Funeral Home was totally covered with snow. Only the tennis ball could be distinguished and helped James find the correct vehicle to shovel out. He needed it for transportation to get to work. The "clown" who stuck the tennis ball on the car antenna turned "hero" that morning. The cold months proceeded into a storylike fragrant spring.

One fine Sunday deep into daffodils, we were at Sunday service as usual. I was feeling a little weak and thought it might be time for some lunch when I found myself lying instead of sitting on the church bench with James cradling my head on his knees. Oh, Lord, everyone close to us looked in our direction. What was wrong with me? As I gathered my wits and sat up, we got stares and strange smiles from the other parishioners. I recovered and endured till the end of the service. On the way out, the pastor sent us off with a strange smile and doubly warm handshake. As we sat in the car speculating about the fainting incident, James suddenly turned facing me and blurted, "I think you might be pregnant." I had thought of every other cause including deadly illness and was terribly frightened. Oh, how happy I was to embrace this life-changing explanation of pregnancy.

We found much to talk about on the way home; now we were going to have a real family like Peggy and Joe. After this incident, we immediately confided in Mrs. Peasley. If James's assumption was correct, we hoped our baby would be born in other surroundings, not in the apartment above a funeral home. Enough remarks by other people had made me aware of the meaning "funeral home."

Chapter 13

THE LITTLE MIRACLE

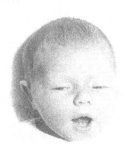

My Daughter Roxanne (1957).

Adapting to my new state in life was not easy. I was happy one moment and the next minute very apprehensive. The constant changes in my body coincided with the seasonal progress of the weather in New Hampshire. There were days when the temperature hit 90°F when living on the third floor became unbearable for me as a German who was used to a moderate climate. When James was at work, I would spend my days shopping and strolling through the pleasant city. Air conditioning was a rare commodity in 1956. Many days and nights that year turned our idyllic love and life into a personal weather hell. I could hardly wait for the fall. Meanwhile, I kept cool walking from store to store, chatting with shopkeepers and neighbors.

One day, as I did my rounds, there was a man trailing me from one shop to another. I got very nervous, almost afraid when he approached me displaying a camera and a smile on his face. "Please, please," I exclaimed, "why are you following me?"

He pushed the camera, hanging on a strap over his shoulder, toward the back and said, "I am Tom scouting for New Hampshire candidates to enter the Miss America Pageant. May I take your picture and ask your name?"

I wasn't quite sure whether to feel relieved or flattered. My response was swift and made him cringe. "You don't want me. I am five months pregnant!" I exclaimed.

"Sorry, sorry, sorry, ma'am," Tom stuttered and fled.

When I told my husband and friends about this encounter, they just cautioned me and advised next time something like this happens to walk back into a store to seek help. The incident was soon forgotten. Tom never again crossed my path.

Our former landlady, Mrs. Peasley, was a true gem; she was very supportive, even suggesting an obstetrician. She introduced us to one of her friends, Doctor Blood, a former governor of the State of New Hampshire. He took me on as a patient to care for me until our child would be born. And before we moved out, Mrs. Peasley gave us a baby shower in her private living quarters. That may have been the first unusual festivity for this or any funeral home. We remained friends after finding an unfurnished apartment farther out of town. James drove away with another friend and bought furniture for a living room, dining/kitchen, and bedroom.

That this friend, Rene, supposedly was one of the marines who raised the large United States flag at Iwo Jima, I found out much later. He and other colleagues helped to furnish our new place. I just moved into another pleasantly prepared nest. Leaving important matters to James was comfortable and wonderful; he always made good decisions.

Our daughter was born in the Concord Hospital a year after we were married. She arrived on a very crisp winter day. The thermometer showed -45°F outside. My water had broken, one symptom Doctor Blood mentioned as a sign we should definitely call him. Luckily, the car started immediately in spite of the temperature, and when we arrived at the hospital, Doctor Blood was there already to take me in his care. He was a marvelous man, always kind and concerned. To keep me from having too much pain, he administered enough medication to knock me out, and I gave birth without the usual pushes and moans. My child was a beautiful, healthy 7 1/2 lbs. baby girl. As customary in those days, Doctor Blood prescribed as her nourishment milk from the Ayrshire cows he kept on his farm. We had had a small baptism at St. John's Church with my favorite brother-in-law, Bobby, as godfather. We named our first offspring Roxanne Marie.

Raising a child for us was by "trial and error." We both were quite young; I was twenty and James twenty-four. We had little support. Ma Helen could not take time off from her job, and Louise, had she been able to come and help, was never married and had no experience raising children. On our first day at home from the hospital, I changed Roxanne's diaper every hour on the hour. When she made "peep," she got a bottle. Bath time with James's participating was fun for all three of us. I spent endless hours trying to become the perfect mother without even a guidebook. On the side, I kept house haphazardly. When it was time, as I read in a magazine, to feed Roxanne solid baby food, I put her in her bassinet lying on her back, and I would shovel spoonfuls of vegetables and meat from baby food jars into her mouth until she sneezed it all over herself and me, signaling she had enough. God and nature saved her life.

Chapter 14

ACROSS THE COUNTRY

We had moved from our first flat above the funeral home to a new apartment farther out of town in Concord. It was also upstairs, but this time on the second floor. The landlady was either divorced or widowed with a teenage son named John. She was nothing like the motherly and refined Mrs. Peasley. James needed exercise, bought a net game similar to tennis called badminton, and engaged Johnny into endless matches since I was busy with the baby and a household. I did not cope too well with baby and chores. James frequently had to break up his badminton games to run a rescue mission upstairs in our home. He handled this with a smile or a reassuring word. He was in charge of most of our grocery shopping and helped with the housekeeping. Bath time for baby Roxanne was an extremely pleasant "mom and dad" job.

The weekends and some evenings were social occasions. We would visit or rather invade a colleague's house. Andy and "Mugs," I never knew her given name, had two young boys. The parents were an incredibly hospitable and patient Italian American couple who fed us, entertained, and helped us over the hurdles that came with being newly married and having to care for a first child. How lucky we were to have such wonderful friends who never meddled but helped when asked. We only realized this when we moved away from them.

I had assimilated well to my new country, but there was always a little longing for the home and customs I left behind. When someone introduced me to another German spouse, believing we might have a lot in common, I was more than eager to find out. We connected by telephone/ I packed Roxanne in her baby carriage after James departed for work and went to meet the German girl in a town restaurant.

Hiltrud was waiting for me; she appeared to be a little older than myself, with dark blond hair, a broad forehead, and stubby nose. There was hardly a smile on that frowning face, showing somewhat irregular teeth when she spoke in a mixture of German and American English. I wasn't exactly overwhelmed by this demeanor. I had hoped to speak my mother tongue as I knew it. To make it worse, the strange tales she told did not make her more desirable as a friend. Her husband allegedly got mad often and would strike her. I could not believe my ears, wanted to run away, and never see her again but could not manage to break off the friendship immediately; I felt sorry for her. I was brought up to ease other people's problems if it was possible and in my power. James, however, wanted me to stay away from the situation. He was afraid I might get pulled into something unmanageable. I never met or even tried to meet Hiltrud's husband, and finally our transfer to Ohio absolved me from my wrongly perceived obligation to right every wrong laid in my path.

Roxanne was thriving in spite of the trial-and-error treatment by me, the new mom without input from any experienced helpers. Days and some nights were totally filled with childcare, like changing diapers and soothing a sleepless baby. I was exhausted; romance between James and myself was nonexistent. I felt myself falling into a pit of fatigue and depression. I slept excessively and neglected my household, caring only for my child. My husband spent weekdays at the airport performing his duties as a controller, weekends helping me, and his only diversion was playing badminton with Johnny, the landlady's son.

I had not realized that James was trying to get back on his interrupted educational track. His father died very early; James was fourteen years old at that time. Mother Helen raised six children on her own. Money for college would go to the older boys first; James was the third boy with five siblings. So he had enlisted in the air force right after high school graduation, assuring he would not get drafted into the army as his two older brothers experienced. Enlisting and serving his time with the US Air Force in Europe as an air traffic controller and, at the same time, playing ice hockey on the AF team, James avoided the draft during the Korean War. Military service also earned him veteran's benefits which, in his plans, would later pay for his college education.

After serving four years in Frankfurt, Germany, the first job back in the USA with the government as a civil servant put him in close proximity to Dartmouth University. His thought was to work full time at the Concord Airport and study philosophy and psychology at the closest university.

However, the distance to travel from Concord to Hanover, where Dartmouth was located, would involve more than two hours per attendance.

Low manpower at the CAA facility in Concord made it difficult to switch work hours with his colleagues, so my husband decided to leave New Hampshire. He bid on or applied for a position within the CAA in another state with a larger facility. James was hired and started a new job with a promotion at Wright Patterson AFB located near Dayton, Ohio.

Chapter 15

MIDWEST EDUCATION

My husband made all the moving arrangements to our next station, rented a small house in Fairborn, and finally approached his desired higher education by matriculating at the University of Dayton, approximately a half-hour drive from our future new home. There at the new work facility in Dayton were also more employees with whom he would be able to trade times and schedules to accommodate his university studies.

In the summer of 1957, we packed the car with necessities while the government moved household and furniture to the newly rented house. We and little Roxanne, six months old, took a long trip across the country into a new life in Fairborn, Ohio. What a change of scenery and living conditions we would encounter! The house we were able to find and rent close to his workplace had only two bedrooms, a small living room, one bathroom, and kitchen. We kept telling each other and resolved that these cramped quarters would be temporary.

Time has wings; it carried us swiftly. James started his studies during the day and worked a lot of evening and night shifts. I got to know many of the wives whose husbands changed schedules to help James clear his days to attend the University of Dayton. Up to that time, my life had been pretty simple, but now I suddenly became initiated into other people's problems, suspicions, and conspiracies.

I became just another "Fairborn housewife." One of the women I befriended used to follow her husband around while I watched her kids. Mistrust is a transmittable disease. Our marital bliss and trust were transformed. I remember starting to worry about James's constant absence even though I knew he had more on his plate than most of the other men who worked with him. I also felt lonesome and bored without him, and lacking other entertainment, I started looking for a job and found one at the church where the pastor needed clerical help. Our back-door neighbor, who was childless, agreed to take care of Roxanne while I worked. In Elsie's home, my little one-year-old seemed well-fed and cared for. I only worried a bit about the large German shepherd chained in their backyard. However, the neighbor lady showed me the interaction between child and dog. When Roxanne, who appeared fearless of the huge animal, tiptoed up to the watchdog, he would crouch to get on her level and wag his huge, hairy tail. She reached out to touch his snout, and he would lick her little hand. I saw and believed the babysitter's positive tale and hoped for the best. There were no dog bites experienced as long as Roxanne remained with the childless neighbor.

My position in the church rectory was pleasant and easy, perhaps a little too easy. Whenever there was a financial drive to raise added money for special projects in church and that caused extra work, the good father would hire more people for the time and tasks involved. He was an exemplary and considerate employer. During one of these financial programs, we hired, and I met a very interesting young lady, Sally, who introduced me to many of her former high school friends. I even joined one of their non-academic friendship sororities, "Beta Sigma Phi." It kept me busy and provided entertainment.

In time, Sally introduced us to her fiancé, Bruce. Sally had a beautiful singing voice and prepared for a musical career. One fine day, she and Bruce tied the knot and departed for New York City where Sally was to have singing engagements on the radio with Arthur Godfrey, a well-known entertainer. We lost sight of the couple for years until James was transferred to Atlantic City. New York City where they now lived, came within reach.

Of course, after a while, James and I needed a vacation! On my first journey across the ocean coming to the United States, I had collected many calling cards, among them one read "Vice President"—of a Dairy Farm on Lake Erie. The gentleman had been very pleasant and invited me to visit whenever I was in his area. I thought Ohio definitely was close to Pennsylvania. As James and I were planning to take a few days off together, Lake Erie seemed a great destination. Grandmother Helen was able to visit and watch Roxanne who was about one and a half years old at the time. I talked my husband into a visit with a totally strange family. This might have been customary in Germany, but here in the United States, it had a different aura. James insisted, but, nevertheless, he acquiesced.

We left Roxanne in her grandmother, Ma Helen's capable hands, and drove toward the neighboring state, stopping at a hotel on Lake Erie. Next day, we searched out the address and gentleman's name on the "calling card" given to me at the discussion table on the SS *America*. The telephone number associated with the name did not function, so we decided to take a chance and just drive to the address and say "hello."

A friendly woman answered the door, and behind her, I spied the man who had been part of the debating group on the ship that carried us to New York. The lady seemed more than a little surprised but invited us on a boat ride and for an evening meal with the family. It proved to be an odd visit with many curious moments. Most unfortunately, James, who felt more than uncomfortable, had one drink too many, and we ended up staying in the couple's guest room with my husband a total drunken wreck and embarrassment. He fell over some stones in the garden and had to be carried into the bedroom by one of the couple's grown sons. I could not depart fast enough next morning and learned a painful lesson. We canceled our reservations at Lake Erie lodge and drove straight back home to Fairborn.

The other calling cards gathered on the ship in 1955 wandered into the garbage can. I was not going to repeat the same disaster visiting another mere acquaintance. I made a mental note: when somebody in this new country casually invited you, they did not necessarily expect you to show up. German customs were different; you only invited whomever you actually expected to take advantage of such a gesture.

Back at our routine in Fairborn, we made plans to buy a larger house and move as soon as possible. Those plans were postponed when we realized I was expecting our second child. My clerical job was not very strenuous, so I continued my work at the church rectory until the day before I gave birth to a son. The pastor was a dream employer. He was absent, frequently looking after the needs of his parishioners. I was a lousy employee who, being pregnant and tired frequently, discovered the sitting room next to my little office harbored a couch. I took advantage of both, the pastor's absence and the unoccupied guest room, to take some afternoon naps.

Until one day, he returned early, tried to enter his home office during my afternoon nap. I did not hear him because I was deeply asleep and had double-secured the doors. A maintenance man broke the locks so they could gain entrance to the rectory and my employer's own home. I was mortified, expected I might get fired, but luck was with me; the good pastor forgave and forgot.

The day of my son's birth arrived. My water broke without much pain or warning, and as happened once before, the doctor I had engaged to deliver my child told us to meet him at the Xenia Hospital.

When we arrived there, a wheelchair awaited me at the curb. I was hastily pushed into the operating room where my doctor had arrived before me. I had none of the expected excruciating labor pains, and when James got back from parking our car, the baby's head was "crowning." Very soon after that, our son, Will, was born, and he was placed on my chest. I thought, *Oh, how long he seemed compared to Roxanne.* And truly, he measured two inches more than my little girl at nineteen inches' length. Then a nurse cradled him in my arms, and he seemed as tiny as Roxanne had been two and a half years ago. He looked up at me with his dark eyes, his head covered with a thick mop of raven hair. I had not lost consciousness because he arrived too fast for me to require medication. I knew he was mine even though he looked much more like a little James, and I fell in love again.

My mother, who had not been very heavily involved in my own care as a child, decided she would combine her first visit to the US with a redemption, offering me a year's care of our son. I wanted to resume working at the church and welcomed her generous offer. She booked passage on the SS *Bremen* to arrive in New York when Wilhelm Thomas, as we named him after my and James's father, was about three weeks old.

My Son Baptismal name Wilhelm Thomas (1959).

We packed both children into our spacious Plymouth and left for the Big Apple to meet my mom as the ship docked. After staying in a motel somewhere along the highway, we drove to the New York Harbor to pick up my mom. We added my mother to the two children in the back seat of our Plymouth which my husband had bought for me to get a driver's license before starting a job. It had buttons to start the engine and was at that time a "state-of-the-art" automobile without a clutch, made for the technically inapt like me.

Roxanne jumped about on the back seat like a squirrel, and the baby was in a car bed without restraints. God knows how we ever made it to New York and back to Ohio without broken limbs or any other accident. We even spent time sightseeing in Washington, DC, on the way home. My mother got to marvel at a gathering of "the American Nazi" club or party wearing armbands with swastikas. She could not believe her eyes. We tried to explain it all as "freedom of speech."

As we pulled up on the small rental house in Fairborn, my mom exclaimed in German, "Where is your house?" She seemed shocked by the size of her home for a year. There was a lot to get accustomed to, but she never complained; she put all her efforts into caring for a two-year-old Roxanne and our infant son under tremendously difficult circumstances. I am sure when she died in 1992 and appeared at the Pearly Gates, Saint Peter took her in personally saying, "You have been through more than purgatory. Come right up, you deserve the highest honor."

Here in the US, living for a year in that tiny house, cooking for us, and helping raise both children for a year, she lost almost 50 lbs. of weight. The summer heat must have been plain hell for her; we had no air conditioning. She persevered and never complained. Only now, looking back, do I realize the trials my poor mother endured after a "diamond" start in life. Born as an only child of a very rich family, WWII robbed her of most worldly possessions and any status. She never looked back, adored her grandchildren, especially our children who gave her the opportunity to shine and show how capable she actually was as a mother when challenged. When my mom went back to Germany after one year with my family, I quit my clerical job and again became a stay-at-home housewife. James pursued work and studies.

On weekends, when most friends had their husbands at home and mine was at work because of his schedule changes that made it possible to attend university, I would seek out my German friends in Yellow Springs who always welcomed me without questions. The Germans I visited on weekends were connected with "Operation Paperclip" started with Wernher von Braun who had worked on the V2 Rocket developed at the end of WWII. He was brought to America to assist in the United States defense program. Many of the "Operation Paperclip" scientists in Ohio worked in the laboratories at Wright Patterson Air Force Base and lived in or near Yellow Springs. Liz E. from my home state of Silesia, one of the scientists' wives, was my first contact to the German gatherings at the Yellow Springs pool. From there, the new friendships developed into a refuge and sojourn with my friends at that homeland community who generously gave of their space and time.

When my mother went back to her home in Germany, our lives began to pulsate in a more normal fashion. We had been very restrained in everything for the entire year. The small house with poor insulation offered no privacy for anyone. Our personal lives were hampered. Romance took another rest. I am surprised we were able to keep even the smallest flame alive. It seemed like a new beginning after dammed-up passions were bringing us close again.

We gathered energy through blissful nights. It seemed like we had spent our days away from each other banned in waiting rooms. Plans for a move to a larger house resurfaced. Time that had been slowed took flight again.

My social life also seemed to resurface again. One day, as I sat in my doctor's waiting room and chatted with the other patients; a lady spoke up to ask about my accent. When she heard it was German, we suddenly had a connection. I had to meet her daughter-in law, Rosi, who was also from my home country. My experience with another German in New Hampshire lay far behind me; I was a bit homesick and ready to find another German friend. I made a phone call a few days later and then stood in front of my house awaiting my future companion. Here she was, Rosalinde, her dark hair nicely coiffed, she was well-dressed, wearing high heels, and was pushing a "pram." *Lovely*, I thought, my kind of German, and in my mind, we were already friends forever. And so, it was! In the baby carriage was Paul, Rosi, and Jack's firstborn.

As we had resolved, James scoured the real-estate market and found a very suitable three-bedroom rambler with a basement in Enon where one of his more congenial colleagues, Ned, lived. We finally bought a more fitting and appropriate house. Our new friends, Rosi and Jack, lived a short car ride away in Fairborn. Summers were hot, and we still lived without air conditioning. We all joined the same swim club and stayed connected through easy and difficult times.

I had not been back in my home country for almost five years because circumstances and finances did not allow it. Roxanne was about four and a half and Wilhelm not quite two years old when we decided I should visit my family in Germany. Travel was expensive, but we managed one ticket for me with a 10-percent charge for an infant under two years of age. Mother Helen, who had not spent too much time with her first granddaughter, offered to introduce Roxanne to the Boston family. In the meantime, Jack and Rosi were going to help out caring for Roxanne and keep James company for the rest of my time away from the US.

William and I arrived in Lingen/Germany feeling instantly at home. My mother called her precious grandson "sweetie pie," and to all the neighbors, he was "the little Willi." Later, after his second birthday in May, while we were with my parents, my son earned another name. He became "Mr. No" since that seemed his favorite word. He distinguished himself in many ways, shutting my mother in her own house by hurling the keys for the front door, which was always locked, out of the window. In those days, you had to use a key opening the front door from the inside or outside. Those were also the days when you polished shoes. "Mr. No" cleaned his grandfather's car seats with black shoe polish meant for shoes only. Willi was an all-round two-year-old boy menace, but the other kids loved him and the change of pace he brought to the neighborhood.

Our vacation time in the spring of 1961 with my German family was over too soon. I had breathed fresh home atmosphere for almost three months, and Will got an early start in the sounds of the German language. My dad loaded both of us into his car. We drove to board a plane in Bremen for our return to the United States.

Chapter 16

DARKENED SKIES

As we approached for a landing in Boston, Will was on my lap, but there was an empty seat next to me, and I thought I may put that squirming toddler in there for safety. It had been a long flight, and even though I was young and energetic, holding a sleeping child was cumbersome to put it mildly. It wasn't just the weight, Will was like a little heater keeping me in constant perspiration. He also tugged on me constantly, either on my hair or my clothes, whatever came in his reach. I would have sworn he had more than two hands. Now that I had him next to me held firmly by a seat belt, I got uneasy, worrying he might squirm out and fall.

Just then, I heard the distinct thump indicating the landing gear was being lowered. I also saw something large and dark flying by my window. Someone behind me said that a tire had blown out on takeoff, but the decision was made by the pilot to fly to the destination, Boston, and face the possible implications there with one single landing. And the piece I saw flying by my window now was apparently some remnant that was wedged in when the landing gear was retracted. We experienced an extra hold in the airspace over Boston, and we were told that we would be landing with fire engines welcoming our flight. I had not flown often enough to really be concerned; my little guy was oblivious to any danger. Will was happily as good as gold. Our pilot was well-trained; his public announcements oozed confidence. Yes, aside from some concern and raw nerves, we landed without further problems; the rescue vehicles departed, and we deplaned.

Welcoming us back at the gate were James and his friend, Dennis, both looking very dapper in fashionable hats. We picked up Roxanne in Waltham and made our way back to Enon, Ohio.

During the next two years, we spent the summers at the Greenacres pool. Roxanne and Will learned how to swim and played with friends. James and I passed the time with Rosi and Jack. Roxanne was a very cute child but always seemed more serious than her peers; she was now getting ready to enter first grade at Mary, Help of Christians Church School. The girls wore the usual parochial uniforms, blue frocks, white blouses with a small necktie. The blue beanie perched on Roxanne's blonde wispy hair made her look so grown-up.

A bus picked the private-school children up and brought them home in the afternoon. Roxanne was a very obedient child and learned very easily. There were never any complaints about her social or academic progress. William, then about four years old, would join the older kids in their exploits around the neighborhood. They spent their time mostly outdoors, unsupervised, reappearing only when they were hungry or in trouble.

One afternoon, we had a swarm of youngsters of various ages running through the house from the back door exiting hastily through the front. My son followed closely on the heels of one the wild older boys who banged the storm door shut behind himself. Will, following the other youngster, stretched out his arms and hit the storm door hard enough to break the glass, cutting his arm and face on the way out.

There was blood everywhere. I arrived just in time to gather my son in my arms. Nothing but total panic ruled while the older boy ran to his house, returning with his brother and their father. They acted fast, loaded me—holding my son—into the car and drove us straight to the hospital a few miles away. Will was immediately admitted into surgery where he received care and thirty-eight stitches in chin and arm. By the time James arrived, our child was doing well enough to be released into our care and allowed to leave with us. We were so lucky to have great neighbors who were at home when the accident happened and quickly came to the rescue; James had to be called home from work, and I just fell apart.

One of my friendlier neighbor ladies used to invite me for lunch when all the children were in school. I would bring a nice bottle of wine, and Annie would make sandwiches. We had a special friendship until a very strange occurrence put an end to our lovely meetings. One day, we were chatting, eating, and drinking our glass of wine when a car pulled into the driveway. Annie jumped off her chair, grabbed the wine bottle from the table, ran into the kitchen, and poured all that lovely grape liquid into the sink, hastily hiding the bottle in the garbage container. I sat there speechless and gasping. In came the husband spying the forgotten wine glasses still half full on the table. He did greet me but looked extremely agitated. I did not know what to make of this situation, wishing I could melt into the wallpaper. My friend had vanished, and the husband sat down to present me with an explanation. It seemed his wife had problems with alcohol and was not supposed to drink but had used my ignorance to her advantage. My luncheon visits ceased. Luckily, we stayed friends.

Raising a son was a challenge. Roxanne on the other hand did not require special handling; she was self-sufficient, played quietly with her toys, and only got boisterous fighting with her brother. Their main disagreements were usually about something silly that she did not consider proper or safe. She always had a motherly streak not quite appreciated by her sibling. I intervened only when William complained loudly, otherwise Roxanne's oversight was a definite help to me and a barrier keeping my son from further injuries. There was one exception. Will was playing in the garage on the electric door. Trying to prevent him from crawling up on the handles, Roxanne hit the electric opener. Will was pulled up, let go when he got too high, and fell on the concrete floor of the garage, inflicting a nasty gash in his chin. Again, there was blood spurting all over. This time, I kept a little calmer and rushed him to our family doctor just around the corner. He received five stitches in his chin, but I did not fear he would bleed to death. Time and age calmed my little devil, but we decided not to take chances enrolling him in public rather than unforgiving private school.

Roxanne gave me a chance to use some of my practical talents, like sewing. Though she usually had the prettiest dresses sent by Louise or other generous family members, I tried to sew her special clothes that made her stand out in the crowd of peers. Her First Communion in church was near, and nobody had sent a white dress for the occasion. What we saw in the shops—I should add, what we could afford at that time—I determined not good enough for our little princess. The weather was still a bit raw as the special Sunday approached, so I bought some white woolen material, designed a dress with a drop waist, and spent days and a few nights finishing a stylish First Communion dress. Roxanne put it on and broke out in tears. "It is so itchy, Mom," she cried. But it was too late to find and buy a replacement; she wore the uncomfortable dress under major protest to that church ceremony. I suffered with her in empathy and vowed to be more selective. I felt just glad and lucky she wasn't obstinate enough to refuse to wear my work of scratchy but fashionable needlework.

As winter approached, my father announced a visit bringing along extra skis to introduce our friends in Ohio to a new sport and pleasant pastime. It so happened that Ned, one of James's colleagues, had married into a local Enon farm family whose land included a suitable hill for sledding and skiing. So my father, who once in his colorful life was a ski instructor, gave first lessons in the US to our friends. He then took back orders for ski equipment to be shipped to Ohio after his return to Europe. We started taking vacations in winter, driving to New Hampshire resorts like Conway and Wildcat. The children, approximately five and three respectively, were fast and enthusiastic learners. Both owned skis matched to their size. We had lots of fun bringing along friends. Our social life seemed on an upswing, but the heavy schedule of work and study still loomed for my husband.

A new university, Wright State, was being built close to Enon. It opened its doors as a branch of Miami and Ohio University in 1964. James thought it would be good to keep my mind honed. We decided I would use my free time when Roxanne was in elementary school and Will in kindergarten to attend some courses. I matriculated with Miami University but attended the newly built "Wright State" campus. Getting college credits for my German high school accomplishments, I started with an English and a history course. I had no idea how much I missed intellectual stimulation and, therefore, threw myself into studies and never accepted less than As. I loved the professors, and they returned the favor with great support and acknowledgment. But as in most human behavior, I soon wanted to expand my newfound freedom and capacity.

We all had received Rosignol skis, and I joined the Dayton Ski Club, becoming a Rosignol representative.

The new entertainment and extra income selling skis made me happy. My German friend, Rosi, watched my children when needed. A new dimension opened that would impact more heavily on our world than either of us had imagined. There were more men than women associated with the sport of skiing it seemed. I met and was exposed to temptations never expected or steeled against. James's work and study schedule was extensive. He never had time and did not care to join me. I took advantage of the time and of my friend, Rosi. I forged business deals and personal friendships in this new social environment. I had not been taught to divorce one from the other. Soon trouble was brewing because of misunderstandings and lack of time and attention on my husband's part. James grew further apart from me as his graduation came closer. We barely spoke. I felt alone at home. James seemed to move in another circle.

One of my customers and friends expanded his interest in skiing into a closer relationship with me. I did not object. My new male friend and I enjoyed other common interests like music. I was somewhat naive, believing that there could be casual, friendly encounters with the opposite sex and simply craved the attention and companionship lacking at home.

One such entertaining evening, my new "friend" mentioned that he had a wife. "Oh, *quel malheur*," I thought. I did not anticipate this. There were many questions on my part like "why are you telling me this now?"

My friend told me bluntly that he was getting a divorce and asked whether I was willing to leave my husband. That same evening, I saw a strange woman sitting at the bar who was in deep conversation with another man.

Francis pointed to her and told me that this was his wife, but he would not introduce us; she was "viscous" and "unreasonable," he claimed. This was a revelation not really expected at that time. I needed to think about the consequences. I gathered my wits and departed for home.

A few days later toward evening as we had finished dinner, James being at home for a change, there was a knock on our front door. The door had small rectangular decorative windows that allowed a view of the visitor who resembled the woman I saw that fateful evening in the club. I must have looked mortified, and James either sensed my terror or he knew more than he let on. "Go in the bedroom." he told me quietly, "I'll handle this."

I slipped away like a thief caught "in flagrante." The visit lasted an eternity, then there was a huge thud as the front door was being shut.

The kids were outside playing, and my husband walked into our bedroom. I suddenly became a third child who, expecting the worst, was totally crushed and worried about what was going to happen. A thousand different feelings overcame me, among them fear of the unknown. James sat down next to me and began in an emotionless tone. "I have known about this escapade for a couple of weeks. Don't have any idea how we got to this disastrous point. Now we have to talk and make decisions."

I could not help but add relief to all the feelings that overwhelmed me all at once. James told me for the first time that he was struggling with personal problems. He was going deaf slowly but unavoidably; therefore, he counted on his study of psychology to forge a new career path. He was aware of his behavior toward me seeming like neglect but did not know how else to handle the situation. He asked straightforward whether I would prefer to leave him or whether we could stay together and try to mend our differences.

After his chivalrous behavior toward me minutes ago, I made my decision instantly without hesitation. "I am staying, and we shall try to attack your medical problem immediately." He did what he did best, he gave me a kiss on my nose, a warm hug, and I melted into his forgiving arms. Later that night, the children safely asleep, we made up for months of frustration and doubt. We seemed to have embarked on a wonderful second honeymoon.

Unfortunately, there were disturbing calls from my "former friend" whom the "viscous" soon-to-be ex-wife had kicked in the groin which put him into hospital. I felt terrible but not necessarily guilty because I could not have been the single cause for the couple's separation. I never saw him again but heard from another friend that after he departed from Ohio, he opened a successful medical practice in another state and married a very sweet lady years after his divorce from his present wife.

Next morning, after the onslaught, James drove away to the University of Dayton and to his evening shift at the WPAFB Control Tower, returning home at midnight. A day later, we called Miami Valley Hospital in Dayton to inquire about an appointment with an ear, nose, and throat specialist. We met with an ENT, Doctor Soifer. James was examined, and the doctor put a proper name on his illness, namely "otosclerosis," a nasty disease that had just then found a cure. We believed Doctor Soifer was the second surgeon certified to execute a "stapedectomy."

James submitted to this procedure on one of his ears, and when that was totally successful, the doctor operated on the other ear. These very difficult operations coincided with James's 1965 graduation from the University of Dayton taking a degree in psychology and philosophy. But he was able, now with almost perfect hearing, to carry on in his field of aviation. Our marriage was saved, and we made plans to move on to bigger pastures.

Chapter 17

A Great Experiment

James and Sigrun at Jame's Graduation with a B.A in Psychology and Philosophy. (1965).

It wasn't easy to mend broken egos and broken trust. My unusual partner put his entire being into a repair mode. I plodded along listening to his philosophical rationalizations hoping for the return of the old, comfortable feelings that had always accompanied our relationship. Somehow, I was not capable of accepting most of responsibility for our estrangement. I tried to argue that I simply had been looking for another "breadbasket" as I called my unfaithful stint. James did not argue, he rarely did before, during, and after. He said, "Let's put it behind us as I did with my hearing problem. Let's do things differently. I did not work and study to end in a deep hole I did not plan on."

Graduation was coming up, and he had to increase study time considerably. But now, I was at and on his side trying to ease his double burden. The graduation day arrived, and we proudly participated in the hard-earned, well-deserved ceremony. I felt like a huge stone had been lifted from my heart and shoulders. With his university degree earned, James accepted a position at NAFEC, National Aviation Facilities Experimental Center, then located near Atlantic City, New Jersey. It since has changed name and location.

He again traveled ahead to reconnoiter the new area and rented a very pleasant house in Linwood within a short distance to his workplace near Atlantic City and also close to Ocean City with all the wonderful entertainment sea resorts usually have to offer. This was now the same man I met ten years ago who captivated my senses and swept me out of my traditions.

The government stepped up, packed our belongings for transport to the assignment, and then we said farewell to a most hospitable state and many wonderful friends in Ohio. We were off to new responsibilities and adventures. Once in New Jersey, we even bought a small sailboat, and all in the family learned to sail; we "soloed" one by one.

Roxanne and William attended Linwood Public schools, made friends, and we established our medical care. Our new primary physician and his wife just happened to be German. This was heaven for me. We organized special dinner parties at our or their home and lived the good lives.

I transferred my university credits from Wright State to Rutgers University in Camden, New Jersey. I would be taking advantage of the free time when both children were in school to further my loves. At first, I signed up for an English literature course which was fabulous and absolved it with another A in my pocket. I also chose to study advanced German literature.

Oh, my luck and happiness cannot be described when I was given the opportunity to have a wonderful woman professor all to myself. We met at different locations; she would instruct me in her specialties: Goethe and Kafka. The professor's name was Sheema Buehne, some would call her a literary genius who had translated the very difficult treatise by the author Emmerich on Kafka from German into English. The precious time we spent reading and interpreting Goethe's *Faust* was an absolute literary "smorgasbord" for me.

James immersed himself in the new job at the National Aviation Federal Experimental Center. He appeared to thrive and was nominated by the agency to attend special courses at Harvard University. All seemed on track for my husband until he lost his university assignment in Boston to a colleague who was chosen instead to participate in the Harvard studies, as I saw it, in part because of his Native Indian heritage. James felt diminished, unappreciated, and disappointed. But as is said, if a door closes, a window may open! So it happened. The United States Air Force was looking for "one good man" for Third Air Force in London/England to run an aviation program and hired James. We had spent eighteen months with the FAA and many fabulous neighbors and new German friends like Christa and Erich in New Jersey. It was time to move on and up.

James made the thought of another move, especially to Europe, very promising. He always managed to put a positive spin on all our lives' changes. He planned many trips to my old home, Germany. In short, this was also to be an opportunity to keep us connected to our European roots. I was overjoyed, though it ended or postponed my studies of German literature in a setting that one could only dream about, like having a fabulous specialty professor all to myself. The children accepted the inevitable, a loss of friends and stability, and James hoped for an interesting future with greater appreciation of his efforts and capabilities.

In the fall of 1968, my husband departed for London, England, to start a three-year posting with the United States Air Force in London/Ruislip, United Kingdom. Our relationship, James and mine, became virtual. Thank God, his ears were now perfect because all there was available to stay in touch, was the telephone. The toughest day of our separation arrived. We heard on the news that a bomb had been thrown into the American Officers Club in London where James was staying while awaiting more suitable housing close to his new position. Apparently, the attack was carried out by Spanish anarchists opposing Franco's fascist treatment of the separatist Basques. Waiting for a phone call from my husband was absolute torture. The children were too young to grasp the gravity of the news, but when that phone call from their dad finally came, there was relief and tremendous joy for all of us.

During one of our wonderful past Christmas parties in the Linwood neighborhood, we had fallen in love with a puppy looking for a home. Naturally, the children were thrilled to have a pet, so we adopted the dog and named him "Brownie." Brownie was a mixture of Beagle and "who knows what." He was mainly a wild animal that did not accept boundaries. He dug his way out from any and all fences. He would vanish and later return smelling abominable having spent the day in the stinky marshes bordering on our house. Or he would be brought back by a police escort. Now that we were moving, a fitting home had to be found for him because there was a six-month quarantine for pets to be expected in England. The expense would have been prohibitive. We advertised in the local New Jersey paper. A nice family showed up, and the children had to say goodbye to their newly found canine friend. Meanwhile, we waited for the air force to move us and the entire household to a home in England and also to be reunited with our husband and father.

A few months passed. Finally, our household was magically gathered by a moving company and transported across the ocean while the children and I flew to Germany to spend some time with my family until James secured a house in Ruislip where his office was located. I heard only good reports from him about his new role and assignment. He sounded like a new man on an interesting mission, being very busy, liking his colleges, superiors, and responsibilities. I could practically see him smiling over the phone lines, very different from the stoic, sometimes too serious man who departed the United States for his air force employment as a civilian.

It took three months to finish building the house we would finally occupy in London. Meanwhile, the government granted us a full-time stay in a beautiful hotel suite in the Royal Lancaster overlooking Hyde Park until we could move into our new home.

I had a special mission in London related to my father. During the war years, he operated his textile business in the Sudetenland, territory formerly Czech, occupied by Germany in March 1939 under the Hitler regime. An acquaintance in the town where our textile business was located begged Dad to extract her son from a concentration camp. And so Doctor F was found, categorized as an essential worker in the war effort, and let out of confinement to join our textile company's management team. He spoke German, Czech, and also English and promised to be a big help in my father's plans and business. He was connected to his English roots. When we finally met his London contact, Anthony, in 1971, we learned a lot!

Doctor F had been an English spy during WWII, and somehow, my father who had turned against the Nazi Party in the early thirties, was aware of his activities. My dad believed that this war could not be won by Germany, and, therefore, he made plans to cooperate with his English employee and possibly save himself, his family, and part of his business fortune.

Chapter 18

SPANNING THE ENGLISH CHANNEL

The modern and pleasant house on Chichester Avenue was finished. We were part of a wonderful Air Force family, trying to adhere to all the customs that come with that new privilege. I organized a housewarming party with the help of other spouses who belonged to the pool of American Air Force dependents. My list of wives was extensive. As I was addressing invitations, helped by friendly volunteers, I was advised not to expect the wife of the current IG (inspector general). *She never participates, is too busy for this, and I'd waste my stamps. Ha,* I thought. *What do I have to lose and addressed an invitation to that lady anyhow.*

The day of my party arrived. We decorated the house, brought in and prepared food, and chose wines and other drinks. As I stood there counting a substantial throng of happily chattering women, the front door opened, and a beautiful person entered with all eyes suddenly gazing in her direction.

"I'll be," whispered someone close to me. I, as the hostess, scrambled toward the door and stretched out my hand to welcome this new arrival. She was beaming and very jovial. I felt like we had met before. It was an instant chemical reaction.

"So glad to meet you," I managed to say.

She grabbed my hand and exclaimed, "I am Fran. Heard a lot about you and wanted to see for myself."

There she stood, the one who would "never" come to my housewarming party. We were instant friends. She was the angel who made my stay in London a pure delight. Of course, James was the other heavenly planner who outdid himself with fabulous trips for the whole family. He made sure we spent time on the continent, visiting my relatives and on holidays like Thanksgiving. We would travel to different cities on the European continent. One such Thanksgiving, we were in Rome, Italy, in the Cesare Hotel. We were told that the Italians did not usually celebrate that holiday, but the kitchen would be happy to prepare a feast for however many Americans wanted to participate. So James posted a list to all who might be interested in a Thanksgiving meal. Over thirty people signed up.

In the afternoon of the holiday, we ambled into the hotel dining room to find every table graced with a red-and-a white wine bottle. We seated the children, ranging in age from eleven to fifteen, at separate tables and ourselves with another friendly-looking American couple. The first course was a ravioli dish and a fresh green salad. Then the kitchen door opened, and one by one, a parade of waiters wheeled in scrumptiously browned turkeys—no vegetables and no potatoes as we would be accustomed to seeing in the American home. We were served only the white meat, and the rest of the birds were taken back to the kitchen. Following the main dish, the waiters served a lovely rum-soaked dessert. And suddenly we remembered our offspring who had been placed four at a table with the children having access to the wines. Too late, the wine had been consumed. The rum-infused cake did the rest. At least our eleven-year-old son was totally wasted.

Then there were those spontaneous short trips from England to Germany. One in particular tickles my memory. It was summer, both children were visiting with my family in Northern Germany when James decided we should have a mini vacation on Sylt, *the* vacation island on the Nordsee in Germany.

We arrived there very late at night. Since this was not planned, we had made no reservations. Driving from hotel to pension and B&B, we found no room at any Inn. James finally had the splendid idea to drive to the beach and spend the short night in the car. Of course, that did not go smoothly. It was a balmy, starless night. We were extremely uncomfortable in the sporty coupe.

So in discussing, weighing our options, we decided to put on bathing suits, grab our towels, and blankets to find a little cove in the sand and spend the night on the beach. Oh, never a dull moment with my chosen daring American love. We settled in; sleep came easily after an exhausting evening. When the sun opened our sleepy eyes, we crawled out of our cove to survey the situation. And what a situation it was: to my horror and my husband's seeming delight, we observed a few nude figures walking toward the water. I saw a familiar twinkle in his smiling Irish eyes. We had spent the night on the "nude" beach. He never told me whether he knew that all along.

James simply slipped out of his swimming trunks and joined the "natives" after trying extremely hard and fruitlessly to encourage me to do the same. I eventually slung a towel around myself and followed him to the car. I was sighing with relief. James enjoyed every moment of my embarrassment. I think he was always hoping I would shake my Puritan upbringing sooner or later. We hastily threw on our clothes and departed to find some breakfast in the village. Still not finding reservations, the "near nude" experience faded to the back of my mind. Though my little devil, James, was a bit disappointed, we decided to drive back to my parents' house in Lingen. There was always entertainment and good food.

Roxanne and William were enjoying the summer learning the German language. My sister, Krista, and family were part of a carnival dynasty running diverse businesses like ghost houses and beer tents in Northern Germany. William, our son, volunteered to sell tickets at the "Geisterbahn," the ghost house, then spent the rest of the time with people whose schooling rarely involved foreign languages and, therefore, forced him to acquire German to be able to communicate. Roxanne stayed with my parents and practiced her "new" language with the neighborhood children. It was an adventurous world for all of us. We made the most of seeing family and making friends.

Later that fall in the UK, we dove again into a pleasant life with work for James as a civilian for the United States and British military. The children were registered at the American schools in London, and I continued as wife and mother taking advantage of all entertainment available to us. Spouses of the military were not allowed to work in the British economy.

I had gained an unexpected friend at our Ruislip housewarming party. The lady, from that day my best friend, Fran, who allegedly "never" came to such parties, did show up, and we instantly liked each other. Lucky me, I now was in the company of an expert who knew the best shops for everything. She knew about "Dior" where we found fabulous cloth to be sewn into regal wardrobes and spa salons to keep us healthy and stylish. Bond Street also had a special store with fashionable dresses my seamstress would have been unable to sew.

"Lillywhites," a department store, sold gorgeous soft cashmere twin sets to match our favorite "must have" piece of wardrobe at that time, our kilts named after different Scottish "clans." Among my wardrobe were "Lindsey," a maroon-and-heather-colored tartan and "black watch" which was actually green and black. And we shopped in that chic store, "Liberty's of London," where you found the most beautiful and unusually designed scarves.

Fortnum and Mason sold every possible kind of tea and also served marvelous lunch. In the Bowles of downtown London, we were introduced to the "Silver Vaults" where James and I purchased a few fabulous pieces like our silver Georgian candlesticks. Other antique stores and auction houses abounded. Aside from all the lunching at Harrods, Selfridges, and shopping at Marks & Spencer, there was that spa near Oxford Street where we could unwind with a

Swedish massage. It was an existence of luxury and leisure one could easily get used to.

Of course, life's little jolts happened as well. We had spent just months enjoying our rented house on Chichester Ave. in Ruislip when the owners decided they needed the home for themselves. It happened to many of our friends, so we deduced that there were some advantages for the English hosts in short rentals to the American military.

Now, in 1970, for the first time in her life, James's mother saw an opportunity to follow the Connerney or Norton roots. The long journey began in London. Her enthusiasm was infectious; James planned a visit to the Irish Republic. Even Roxanne and William got excited. We piled into our Grand Prix and crossed by ferry from Holyhead to Dublin. The first order was to find members of the Norton clan who hailed from county Galway. We found them in Tuam. As we asked for directions, the local fellow sent us proudly toward a "two-story" house. Anticipation grew until we finally arrived there without notice. Nobody had been in touch for years.

Three people emerged from the home toward our car, spilling its four occupants. My first impression as smiles and greetings explained our visit. I counted five teeth among three ancestors. We were escorted into the parlor, a spacious sunroom with a dirt floor. The tea was prepared—mind you—in the hearth. But the Porzellan cups seemed very fine and precious.

One of the women tugged on Ma Helen's sleeve, exclaiming, "Come, have to show you my beautiful ass." Helen, totally taken aback, stumbled along with the children trailing wide-eyed, now gazing upon a harnessed donkey pulling a cart of dark dirt. Peat farming was the family enterprise. As far as our eyes could see, there were miles of bogs. Ma Helen sent care packages for years!

We left our ancestors behind and enjoyed a meal of fresh salmon at the inn. So fresh, we watched the fish jumping into the nets as they swam upstream to spawn.

The next day belonged to the Connerneys. Helen's ancestry began in Cork at the Ring of Kerry. We stopped at Blarney Castle to "kiss the Blarney Stone." It became quite obvious as I bent down backward to find the "Blarney stone" why those chosen in the Middle Ages for this torture developed angel tongues. Whatever the people who bent their back to kiss the object over a frightening abyss asked came tumbling out of the victim's mouths willingly and immediately. One slip of the holding hand and you met eternity. These days, there was a "saving grate."

We could not depart fast enough. Ma Helen had no addresses. The Os, like in O'Malley or Mac's, dominated the Cork name register. Our former Connerney gladly came away to explore Killarney and the rest of the Ring of Kerry.

Well, a year before, 1969, as James and I were first exploring ancestral Ireland, Neil Armstrong and Buzz Aldrin had landed on the moon. Now a year later, Ma Helen finally made it to Ireland.

The rental price for the Ruislip home suddenly shot high above the allotment we received. We looked for another house in the London area and found one in Northwood which had been occupied by a departing American dentist and his family. It was a bit farther away from James's place of work, located in the north of London, but was directly on the "tube," the mass transit system serving all of London. Our future residence was a large, older home with a lot of history. It was called "Paramata" after the first settlement of British prisoners in Australia. For the children, it presented some unusually spooky aspects. The walkway to the house was paved with ancient cemetery stones, names, and dates still distinguishable. James and I loved this spacious home in Tudor style with authentic Norfolk locks on all doors, a fabulous English garden, and grand neighbor like Mr. Forbes, a widower who had a rather regal and very protective housekeeper. Mr. Forbes, as he told us, owned a tea plantation somewhere in Ceylon, and he recited wonderful exotic stories.

I was in heaven again, but William, our ten-year-old, developed fear-related physical symptoms in part because of the hectic and late-night social life James and I lead at the American Officer's Club. We thought the children were old enough to be left alone in the late evening. Roxanne was thirteen and William eleven years old. We did not know and consider that the entertainment offered in the evenings on local TV, compared to our former TV presentations in the US, consisted mostly of ballroom dancing. I wonder whether that experience contributed to a later dislike both of our offspring displayed against terpsichorean exercises.

From my new friend, Fran, I learned new forms of artistic expressions like brass rubbing in the surrounding churches. One early morning on a cooler than usual day, I picked Fran up, and we started toward Hillingdon to tackle brass rubbing. There were a number of sarcophagi holding the remains of knights and their ladies as here the "L'Estrange Memorial." We placed our special black paper on the likenesses of John and his Lady Jane to trace the outlines with our gold crayons. We spent almost six hours at this task. It was hideously cold in the church, and yet we persevered leaving finally with painful limbs courting sniffles.

There was nothing she did not know about London, its beauty and opportunities. I felt like an angel had been sent to guide me through an amazing new world while our husbands kept our lives protected and safe. We did not meet Fran's other half, Chick, for a long time. He was mostly busy and traveling in his job as IG, inspecting American bases or whatever an inspector general's duties entailed. We never saw them at the officers' club on any of the evenings we spent frolicking with other friends. We actually met Charles or "Chick" one night when all four of us attended a new play on the London stage called *Abelard and Heloise* written by an author called Milar. My strongest memory of that occasion was embarrassment about the very open sexual content and display on stage. I was back to being the Ursuline student, exhibiting the demeanor formed by my prudish education in high school. Chick, being extremely observant, noticed and grinned, amused. I tried to avert my eyes and ended up looking straight into his. There was a twinkle of understanding and a familiarizing kind grin that made him my ally and friend until his passing at age ninety-three. Fran, in her extraordinary fashion, just laughed everything off and restored reality.

One fine day after we had moved into the charming North London house, Paramata, the doorbell rang. When I answered, there stood a young man in a long dark blue military coat, his shoulder-length brown hair somewhat tousled. He had a delightful little smile on his face and carried an old doctor's bag in one hand. Looking down on me, he began, "Remember me, Uli, your cousin from Münster/ Germany?" Of course, I did; he was my uncle Eduard's youngest son, one of eight children. He was also one of my favorite cousins.

"What are you doing here? You didn't call. What happened?" I clamored.

"My mother died unexpectedly, and I just needed to get away," he said with a very sad face. *Oh, dear, what next*, I thought and gave him a loving and welcoming hug. Here was a somewhat puzzling but demanding family responsibility rearing its head. When James came home from work, we decided together that Uli was welcome and should stay as long as he felt at home with our family. My husband was always ready to lend a helping hand; again, he proved to be the harbor in a storm. After a few days, I determined that Uli's hair needed to be cut so he would fit in with our military family. I took him to Luigi's, my hairdresser, who adjusted his locks to go with our lifestyle. Uli did not object then, and I actually had no idea that he, not unlike Samson, considered this act an invasion of his masculinity as he confessed many years later. But he must have forgiven me because from the day Uli showed up at our door in London, he became a part of our family, joining us in Washington, DC, even after he absolved medical school in Germany.

In later years when we returned and lived again in the United States, I found him a three-month internship at

George Washington Hospital, and he spent eighteen months with a German research grant in the Commonwealth Hospital, Richmond, Virginia. He repaid our hospitality in kind when he chose Bern, Switzerland, for his first home, starting a career as a budding neurosurgeon.

We had gained another family member when my cousin, Uli, appeared. That was not to be the last surprise addition. The years had brought changes in Offenbach, Germany: my aunt Benedikta succumbed to an illness, and after years of study and teaching, my cousin, Sr. Monika was voted "Mother Superior." One of her teaching "charges" in school was the oldest daughter of an Offenbach patrician family. Karin was to spend time in London, learning the language and needed a "home away from home" for the weekends. So we temporally adopted another daughter of Roxanne's age. As her English grew, so did our bond. We departed for High Wycombe as Karin finalized her year in an English boarding school and returned to Germany. We remain connected to this day.

Time ran out at Kewferry Dr. in London. The rent again was hiked beyond our expectations, and we decided to move into military housing in High Wycombe, best known in America for the famous "Hellfire Club," having had Ben Franklin as a member of an American Assembly Emissary seeking easing from British taxation of the American colonies. We were entitled to military housing, and the air force assigned us a town house close to the home of our American military doctor. Doctor John soon emerged as James's favorite tennis partner, and we met and enjoyed his wife Robin's company during the many wonderful pub crawls undertaken on weekends. We would pile six neighbors in a car and set out to visit pleasant entertaining "pubs," restaurants, and "drinking wells" in the English countryside. It was a good life for all. New assignments brought satisfaction for James. His efforts in designing the airways for the special speed run of the SR71 airplane, New York to London, earned him the Civilian Meritorious Service Medal.

James also graduated from the Air War College while pursuing his private pilot's license.

In early summer of that year, we had a most wonderful experience that turned out to be the highlight of our six-year assignment to Great Britain. We received an invitation to attend a formal tea or garden party given by Her Majesty, Queen Elizabeth II, at Buckingham Palace in London. Of course, there was protocol to be followed. The ladies invited were advised to don hats; military members had to wear their dress uniforms, or if they were civilians, they could appear in "morning suit," a type of tuxedo. The latter, of course, in our case had to be rented. So I set out to Bond Street to buy a special dress and hat. James found that English tuxedo with a top hat and also rented a Bentley with chauffeur for this special occasion.

We and all around us were very excited when the day of the party arrived. Pictures could only be taken "on the Mall," the avenue leading to the palace since no photographers were allowed in Buckingham or at the tea party proper. It turned out to be a grandiose occasion. When our driver dropped us off inside the gates, we were ushered upstairs and loudly announced by a footman as "Mista and

Misses Norton" while he tapped a wooden staff twice on the floor. We then proceeded into a large room and from there back down into the beautifully groomed gardens behind the palace. There were tents with food and drink, mainly small savory sandwiches and sweets. We spied many well-known faces we had only seen in magazines before.

Finally, the Queen arrived. She wore a lovely blue summer dress, a small, brimmed hat topped her well-coiffed brown hair, and she was carrying a purse. That purse somehow impacted and struck me as memorable and odd: why would she need a handbag in her own home? No one ever gave me a good explanation for that. I guess the home or palace is too large to find whatever you need wherever you are? We did not shake the Queen's hand. She walked slowly past us, smiling, asking a question here or there; we smiled back and were instantly transformed into ardent royal fans.

Year 1971 in front of the mall at Buckingham Palace.

Chapter 19

A DIFFERENT REPEAT

That summer of 1971 was such a highlight of our tour in Europe that James decided to sign up for another three years. We both were totally happy and fulfilled with activities; we assumed nothing would change. But that is never the case when you are part of the military establishment or the human race. Reasons were never given; however, Third Air Force moved to Mildenhall/Lakenheath, farther north from London, and our family followed right along. The children changed schools and had to make new friends and adjustments again. The new town house assigned to us seemed a bit crammed, but when you are part of the military, you just "fall in line." I guess "you are serving, not being served."

James suddenly spent a lot more time traveling. Most of the time, I did not know where he went and when he would return. I knew not to ask too many questions. We lived by special rules. Similar to marriage, it is for better or for worse.

It was a British world of strikes of every kind, postal, railroad, all branches of life seemed affected. I used to say France has 365 kinds of cheeses, and Britain in those days had 365 days of different strikes.

We also had diverse neighbors, like a navy couple with younger children. Life became more "Americanized" with neighborhood parties and gossip. I missed London and the old friends fiercely, especially my best friend, Fran. She and Chick had been transferred to Upper Heyford some time ago. Chick was now a wing commander and apparently traveled less.

Our children went on to new schools: Roxanne to an American high school in Lakenheath and Will to a base-connected elementary school. Making new friends is never easy for anyone, so it turned into adventures for both kids. Roxanne found herself attending with some young people who had famous parents, like the daughter of the country singer, Glenn Campbell, and others who later would be in headlines themselves. It was a "heady" and heavy time as I found out from my eleven-year-old son. Apparently, he had to find a "bigger guy" who would protect him from bullies.

James and I were mostly oblivious and not tuned into the new problems enveloping our offspring. We grown-ups were wrestling with disheveled schedules, the annoying mail strikes, and other general changes. Many adjustments were due to our closeness to the American neighbors. Some were disruptive, some just difficult to handle. We had to endure other people's loud bells when they gathered their family at dinnertime, and we felt obliged to attend neighborhood parties that called for reciprocity. And mainly, we coped with frequently absent husbands who were conducting business in a "cold war" setting. My own husband used to bring back rewards and trophies, such as "Labour Chairs" from Spain and a large globus-shaped cocktail bar from Italy which today still is a desired item in our family. We, wives and mothers, took family trips to the continent with our children but without fathers. All of us made the most of the time in Europe.

The year 1974 rolled around too fast. It was time for us to return to the United States. As I was told by other women, the military had a penchant for sending the military spouses on TDY, temporary duty, whenever a family move was in the picture. So it was for us: James departed for temporary duty to the Pentagon in Washington just before Christmas 1974. I was left by myself with the two children to empty out our quarters, a town house, selling whatever could be sold, and cleaning for an end inspection.

Times were difficult. The economy was suffering an energy crisis, 1974–76. It made selling our car extremely hard. I finally sold a beautiful Grand Prix sports model for $100. Everything else was under a clouded star. We needed cash to buy a house back at home and actually had to ask for help from the family. James, now stationed at the Pentagon, had taken up temporary residence in Alexandria, Virginia, provided by the air force and was actively searching for homes to accommodate us when we could proceed from England back to our home country. We finalized our move out, and the children and I caught a space available flight to Washington, DC, arriving just before the holidays. James had bought a home, but we could not take occupancy until the New Year.

We moved into a suite in Alexandria which was close to our future home in Springfield, Virginia, so that I could transport both children to their school. Roxanne became a junior and Will a freshman in West Springfield High School. Thank God they went to the same school since I had to drive them there and home for almost three months.

The entire return to America was like a shock treatment. Of course, others who had preceded us had the right words to describe that "condition." In Europe, we were "big fish in a small pond"; now suddenly, we had to adjust to being "small fish in a big pond." After six years of fabulous life and adventures, it felt like we were hit by a cold shower. The air force tried to make our adjustment as painless as possible. They appointed "sponsors" for every returning family. We were assigned to a wonderful military officer and his family who helped us in every way to adjust and even to find friends. They were gracious and kind with a memorable huge Great Dane as pet. Ordinary life took us, dipped us into a fast-moving stream of work and obligations, and we unfortunately lost touch with our benefactors.

Chapter 20

THE BIG POND

Abstract of Memoirs

We arrived in the big pond
and dropped anchor.
Swim or sink
was the motto then.
We searched the horizon
For helping hands.
We found them in one man,
a steady friend.
You gather your talents,
he told me then.
I bring the magic,
you add the labor.
Together we weave
a basket of life.
That will carry your loved ones
and you in style.
This I did as he wished,
and we sailed with a smile.

The adjustment returning into the "big pond," the United States, was *huge*, even bigger than the one I had to make on my first arrival in 1955. This one involved so much more and more of us. I could not just stretch "all four limbs" and totally encumber James with everything as I had done after I met him. Our lifestyle had always been supported adequately by one income, his alone. When we owned our first home, the house payments and the children were small. The year 1974/5 presented challenges everywhere. The children were advanced enough in school that it impacted their future. Especially Roxanne who was faced with a major choice of college and a field of study. Will, who excelled in tennis, needed to become more comfortable in that competitive sport. There was no place for our son to practice his tennis game. We actually joined a country club to provide tennis courts for practice.

But the main problems concerned the usual evil, money. We had borrowed from the Boston family to buy the house in Northern Virginia thinking James's Pentagon position was safe. However, the general whom he accompanied and for whom he worked was transferred to another stateside base and offered to take him along as in 1975 my husband was still attached to the air force as a civilian employee. But as happened before, fate intervened in our favor. One of James's best friends during his involvement with the SR71 in Europe was now also living in the Washington, DC, area. And the lady who had handled our real-estate needs thought I was "a natural" and was my guide into that profession. Bob, James's friend, was the kind conduit to a wonderful built-in pool of clients and customers departing and arriving in our area. I was very happy to offer up any further literary studies to help my family financially.

As in all well-laid plans, there was one major problem—namely, staying in our house and in Washington, DC. We had made friends everywhere including in the direct neighborhood. Across the street from us lived people with connections in the US government. When I mentioned my distress, I realized that good fortune was on our side. One person happened to be the liaison between the president (of the US) and the senate and was able to find an ear for our debacle. He knew the current head of my husband's former employer. James applied to return to the FAA as a staff member at their headquarters in Washington, DC. He was accepted, and we proceeded with work, school, and home. The "little fish" adjusted well and started new lives in a big, adventurous pond.

Grandmother on my Father's side. Her Name is Clara Groeger.

Grandfather Alfred

Grandmother on my mother's Side Selma Tuckay Kauffmann she was a Pianist of Hungarian Descent.

Married Reinhold Kauffmann, who owned the textile mill.